ECO
BABY

ECO BABY

A Green Guide to Parenting

Sally J. Hall

green books

First published in 2008
by Green Books Ltd, Foxhole,
Dartington, Totnes, Devon TQ9 6EB
edit@greenbooks.co.uk
www.greenbooks.co.uk

Text printed on 100% recycled paper
and bound by TJ International, Padstow, Cornwall

ISBN 978 1 903998 90 8

Contents

For my very own Eco Babies,

Jamie and Lulu

Introduction

"Become the change you seek in the world." – Mahatma Gandhi

The moment when you realise you are expecting a baby is a truly magical one. It is also a time when many of us take a step back and ask what kind of a world we are bringing our baby into.

If you are concerned about the environment and what the planet will be like when your baby is grown up, you have to start making decisions about all areas of your new life as a parent. From nappies to food, from nursery furniture to toiletries, there are green and ethical considerations to be made with every decision and purchase. Climate change is now a fact of life and in addition to the obligations of national and international governments, each of us has a moral responsibility to reduce our carbon output as well as our use of precious natural resources.

Whether you are a keen novice when it comes to environmental issues or are already a fully paid-up eco-warrior, becoming a parent is a totally new ball game, with a bewildering array of consumer items to buy and choices to make. The very nature of the baby industry means that enormous amounts of raw materials, energy and transportation costs are expended annually to make new products for the babies being born in this country alone. We have a culture that places more importance on appearance, fashion and trends than on the need to conserve our precious resources.

This book gives you clear and practical advice on choosing green and environmentally friendly products, tells you all the key facts you need to know and helps you to make the best decisions about each and every one of the products you will buy for your new baby.

It is easy to feel overwhelmed by what is a huge subject. The good news is that every tiny step you take towards a greener, more sustainable and environmentally conscious way of living, the better it will be for our planet and for your child's future; every little bit you do really does make a difference. There's no need to feel that you must change every aspect of your life all at once; changes can be made gradually and then become part of your daily routine. To make going greener in your home a viable proposition, it must be achievable; luckily it is, and the environmental option is often cheaper for you too.

How to use this book

This book covers pregnancy through to the first year. Each chapter can be read on its own, or as part of the whole book. For this reason, some of the facts and information are repeated, so that you can dip into the book as and when it is relevant for you. You can find information on companies who offer green and ethical products and services in the *Resources* section at the back of the book.

This book deals with a number of issues including allergies, intolerances and skin problems, plus suggestions for remedies and treatments. If you have any concerns over your health or that of your baby, you should always discuss these with a doctor or your midwife or health visitor.

The most important thing is to enjoy your new baby!

Sally J. Hall

Commonly used terms

CARBON FOOTPRINT

We hear a lot about reducing our personal carbon footprints. But what is a carbon footprint, how do you calculate yours and, most importantly, how do you reduce it? To put it simply, it is how much carbon dioxide (CO_2) we cause to be added to the atmosphere, and therefore how much global warming we create as an individual. Many aspects of modern life can produce carbon dioxide – from driving and flying to cooking and cleaning or purchasing products that required energy for their production or delivery. Sometimes it is fairly easy to understand how we cause emissions – air and car travel are good examples. But there are plenty of things we do where the consequences are less obvious. All the electricity we use from conventional gas- and coal-fired power stations creates CO_2 in its production. So, for example, we create CO_2 when we use energy in our homes; we also use power in our places of work and in public areas (libraries, street lights, traffic signals, public transport etc.). If we buy goods from abroad the miles those goods have travelled all add to our carbon footprint.

We are all able to help reduce the overall emission of CO_2 in many areas of life. Simple remedies include things that seem rather obvious – walking or cycling instead of driving, turning off lights in unused rooms, turning down the heating, using public transport instead of a car, turning the thermostat down by one degree, insulating the loft, shopping for local, in-season produce instead of choosing food flown halfway around the world and turning off any electrical appliances at the mains when they are not being used. If we want to go further, we can switch our power suppliers to those investing in renewable energy, and investigate other energy-saving technologies, for example wind turbines, ground-source heat pumps or solar panels.

For further information about calculating and reducing your energy consumption (see *Resources*, 1).

ETHICAL

This means that the product has been produced to fit in with the ethics or principles of a particular group. However, within the concept of this book, it is generally understood to mean that the product has not caused any harm to the environment, to people or to animals.

FAIR-TRADE

This refers to a system by which we can buy products that have been grown or made in Third World countries, using a system that ensures that the farmers and workers are paid a fair price for their goods, are paid on time and are assisted with investing profits back into their businesses and communities. Fair-trade schemes often support co-operatives of labourers or artisans who are working for themselves, rather than being employed by others. For the people involved in these schemes, fair-trade is better than aid; it helps the producers to use their skills to build a sustainable future for their families.

Fair-trade labelling now covers a range of produce and much of this is in the food sector. If you are unable to buy local, organic produce, then one that has been fairly traded makes a good ethical choice, especially with items that are not grown in the UK such as bananas, tea and coffee and chocolate. Remember however that by choosing foreign produce you will incur food miles, so weigh up the pros and cons before you buy. Fair-trade produce is not automatically organic – check the labels if you are concerned.

In the clothing sector, it is not always easy to check the credentials of suppliers – clothing labels do not give the full story. If you want to find more information on the ethical and trading practices of shops, there are two good sources of information:

Ethical Trade is an alliance of companies, non-governmental bodies and trade union organisations which aims to promote and improve the implementation of corporate codes of practice which cover supply-chain working conditions. Its goal is to ensure that working conditions are acceptable for all workers making products for the UK market.

Clean Up Fashion is a pressure group which publishes details of many high-street retailers, giving you details of their ethical standards and how they treat their workers.

There is more information on the fair-trade issue, and fair-trade products in Chapter 5 on clothing and in Chapter 6, which looks at food. In addition, you can visit the website of the UK's Fairtrade Foundation at www.fairtrade.org.uk (see *Resources*, 2).

The larger retailers have some way to go to catch up with ethical trade, though some have started to stock organic cotton garments, and some do follow ethical practices. The fact is that as more of us demand fair pay and conditions for workers in Third World countries and vote with our wallets to buy ethically produced products, the better the situation will become for those workers.

FOOD MILES

This refers to how far food has been transported. So, for example, bananas from the Caribbean and mangetout from Africa are transported in refrigerated ships or worse, in planes, halfway around the world. The ease of transportation has led to consumers being offered out-of-season products all the year round – for example strawberries or green beans in winter – and a lack of awareness of seasonality. This has environmental impacts due to the fuel used and therefore the additional carbon dioxide added to the atmosphere, and in many areas of the world it has led to the loss of local economies and damage to the natural environment as areas are cleared and irrigated for cash crops instead of growing subsistence foods for the local population.

GREEN

In this book, I use the word 'green' essentially to refer to lifestyle, ethical and political choices that help to conserve, rather than harm, our planet. The more of us who choose to go down this route, the better we will become at making choices which are sensible, sustainable and nurturing for our world – and that of our babies.

ORGANIC

This refers to the way in which a product is grown without using pesticides and fungicides. The Soil Association is the UK's main ruling body on organic food crops and animal products, with very high accreditation criteria; however, many countries, supermarkets and other distributors have their own organic standards which do not necessarily meet the strict criteria of the Soil Association.

SUSTAINABLE WOOD

The Forest Stewardship Council (FSC)
The FSC is a widely used certification for timber and other tree products; choosing wood or paper products bearing the FSC mark means that you can be confident that the forest source is managed sustainably and does not contribute to global forest destruction.

For more information see www.fsc-uk.org.

Programme for the Endorsement of Forest Certification schemes (PEFC)
A global, independent, non-profit, non-governmental umbrella organisation which promotes sustainably managed forests through independent third-party certification. For more information see www.pefc.org.

If you buy wood or other tree products bearing either of these marks you have the assurance that you are supporting the sustainable management of forests.

Chapter One

Getting ready for your Eco Baby

When you're an expectant or new parent, it is easy to look at lists of things you may need for your baby and feel as if you absolutely have to have everything on them. There is huge pressure on us to have the latest pram, the most fashionable high chair, the sling all the celebrity mums have, and the 'designer' accessories – and you want the best for your baby. So it is little wonder that many of us feel that it is the norm to spend an enormous amount of money on our little ones – and also that everything we buy has to be new.

It's difficult to envision how we can be green when buying for a baby – one step into a baby shop and you are surrounded by a sea of brightly coloured metal and plastic items. Remember that baby lists are only suggestions for things you will need or may find useful, so just use them as a good starting point to find out what is available. Bear in mind that the baby products industry works on the premise that parents are a captive market and may be persuaded to buy new products for each baby, responding to changes in the family's needs and also in fashions. If you can combine all of that information with an eye on the greenest and most ethical choices, you should end up having only what you need and with a light carbon footprint.

In general if you are buying wood items, look for FSC- or PEFC-approved products, and avoid MDF (Medium Density Fibreboard) which is a material that has been used a great deal in recent years, and has been the wood substitute of choice in scores of DIY programmes on the TV. This composite is formed from particles of wood bonded together with glue. Its use has been linked to health problems including lung cancers and allergies. There are two potential hazards: if you cut, sand or carve MDF, the minute wood particles and dust can become lodged in the lungs (It is recommended that you wear a mask and eye protection when working with MDF for this reason); and the bonding resin of the MDF may give off fumes in the form of formaldehyde, which can be carcinogenic.

BUY ONLY WHAT YOU NEED, AS YOU NEED IT

Where possible buy things as you need them, rather than making expensive mistakes or buying too many items that you then have no need for. For example some friends bought an expensive Moses basket before their baby was born with all its sheets, blankets and accessories, only to have a baby who was so long, he was too big for it even at birth! Conversely, my own daughter was tiny and slept in her basket for nine months, at which point it made far more sense to buy a cot bed than a cot, as she could then use it as a bed when she was old enough. So my friends could have saved the price of the basket, and I had no need for a cot.

Prams and buggies are an area where it is easy to get it wrong – the big, three-in-one push-chair which seemed so sensible and durable in the shop may be completely impractical in your flat, the boot of your car or on public transport, whilst the lightweight buggy you thought was so cheap and manoeuvrable may not be at all comfortable for your baby to sleep in.

> ECO TIP: Pick up a second-hand baby bath. Most will have had very little wear. Use second-hand, buy when you need it but not before, and pass it on to a new mum when your baby has grown out of it.

BUY SECOND-HAND

Buying second-hand from friends or family, online or through advertisements ensures that the nursery furniture and other items have a longer life and are not disposed of after they have been used for just one baby. It reduces our consumption of new items, thus reducing the need for more raw materials and energy to make the goods, and the pollution caused by making new products.

WHAT DO YOU REALLY NEED?

It can't be said too often – buy when you need it, ask for hand-me-downs and, above all, recycle when you've used it! The following section is a guide to recommended items and green options – with careful planning and by leaving the purchase of some items until you need them you can cut down on the total bill and your carbon footprint. I have not included clothing and toiletries here, as they are dealt with in separate chapters.

FEEDING AND CHANGING

Bath time

One of the best examples of an area where you may buy unnecessary products is for baby's bath. When babies are tiny, you really only need to top and tail them most

BUYING SECOND-HAND

See before you commit

See the product before you buy if possible, so that you can check for any breaks, snags, tears and missing parts. If you are buying mail order or online, make sure you have the right to return it if there is something wrong.

Make sure all the parts are there

Try to make sure any fitting advice and instruction books are included. If they are not, try contacting the manufacturer of the product to see if they have an instruction book you can buy or download from their website.

Support your local second-hand shop

Second-hand nursery shops are great places to buy from and it pays to get to know the owner and ask for advice. Owners will have seen so many different brands and styles of product, that their advice will be invaluable. If you have a particular product in mind, ask to be put on a list to be called if it comes in, or keep popping in regularly.

Reuse and recycle

When you have finished with things your baby has grown out of, give them a good clean and take them along to the second-hand shop, have a sale with other mums, or simply give them to a pregnant friend or to charity – there are some baby-equipment charities who donate goods to families in need of them. These tend to be locally based, so check your Yellow Pages or ask your Health Visitor.

Reject second-hand goods if you have any doubts at all about an item

More advice

The website of the Child Accident Prevention Trust.[1]

Before you sell

The Baby Products Association has a leaflet for potential sellers of baby equipment called 'Think before you Sell – second-hand can cost lives'. It shows you what to look for to ensure that anything you sell is safe for another baby.[2]

days and give them a full bath a couple of times a week. It is not necessary to give a full bath every day and it can be quite drying for baby's skin. So to spend money on a baby bath may be a waste, especially as they are made of plastic that can be difficult to recycle. Before the invention of plastic and foam, our mothers and grandmothers used to wash us in the bath or even the basin. My daughter was tiny and felt more secure in the bathroom basin than in the bath – and it was much easier on my back! Just be careful you don't burn your baby on the hot tap. Better still, why not share a bath with your baby?

Other bath aids may be useful, depending on your circumstances. If you have twins, little angled bath seats or cradles are useful to keep one baby safe while you wash the other. There are also various pads and supports that keep the baby's head above water, allowing you to use both hands to wash!

The best bath-time product I found was one used from when the baby can sit up – a seat with a bar encircling the waist, usually with some play beads on it – in which the baby can sit and play for ages. While they are plastic, you can pick up a second-hand one and pass it on when you have finished with it. Remember never to leave a baby unattended in the bath.

Bibs

There are bibs of all shapes and sizes available in nursery shops and pharmacies, usually with a daft picture and soppy comment on the front! Bibs are often made with a towelling or cloth front and a plastic backing that prevents wet and soft foods leaking onto the baby's clothes. However plastic-backed bibs tend to become cracked and perish when put in the washing-machine and it is best to avoid plastic wherever possible. A good quality towelling bib without a backing will last longer and wash better.

Avoid the tiny, first-size bibs; buy the larger ones and the whole baby will be covered when very small – and you will appreciate it the first time you try weaning foods like spinach or tomato-based purées! Bibs with an elasticated hole to go over the head are much easier to pop on and off than ones with fiddly ties. Look for organic cotton bibs, some of which are fair-trade items (see *Resources*, 1).

Bottles, teats and soothers

Firstly, it is important to say how fantastic breast-feeding is. Quite apart from the nutritional and emotional benefits to the baby, which are explored in Chapter 6, it is the best environmental option, too. You require very little in the way of equipment (except for a few well-fitting maternity bras of course), milk is instantly available and requires no mixing, sterilising, storing or washing-up. Occasionally you may wish to express your milk and store it for times when you can't be around to feed your baby. If you are able to successfully breast-feed almost exclusively, you will save resources in terms of the bottles, teats, lids, sterilising equipment, energy use and formula milk that would otherwise be manufactured, packaged and transported; you'll also save yourself quite a bit of money.

To avoid using plastic, look for glass bottles (see *Resources*, 2). If you buy plastic bottles be aware of the potential problems: phthalates (pronounced 'thalates'), are substances used to make plastic items softer and more flexible. They may give off harmful chemicals that have been linked with reproductive problems. However, from January 2007, the six most dangerous phthalates were banned from use in all baby items, including bottles, teats, drinking and feeding equipment in the EU and in the UK (though they are still present in some baby toiletries, particularly baby lotion, shampoo and powder – see Chapter 8). So when buying feeding equipment, look for products which have the CE (EU-approved) logo and also look for the British Safety Standards numbers, BSEN 14350 for drinking equipment and BSEN 1400 for soothers.

In addition to concern over phthalates, there have been studies to suggest another substance, bisphenol-A, can be leached from the plastic of baby bottles into their contents and can cause health problems including neural and behavioural problems.[3] This is another reason to avoid plastic bottles and to breast-feed as the first choice, choosing glass, or bisphenol-A-free bottles if you cannot.

Crockery and cutlery

You needn't turn your kitchen into a primary-coloured plastic world – bear in mind that until fairly recently, mums would simply use the cutlery and crockery they already had in the kitchen for their baby. You can feed just as easily from an ordinary teaspoon and use a china bowl for your baby's food – just don't use your heirloom Doulton!

Whilst you can avoid plastic cutlery – a few companies are selling sets made from bamboo (see *Resources*, 3), it is more difficult to avoid other un-recyclable plastic feeding equipment. However, there are a few items that are really useful – a bowl with a lid is great for trips out with baby; a plastic plate will survive being thrown from the high chair; plastic cups are durable, unbreakable and really do last for years. As long as your products were bought after January 2007, you can reuse them for a long time – they are useful for picnics, as pots for painting or for using for younger children when they visit. If you are buying leak-proof weaning cups, choose the larger size ones that you can use as beakers later on, without their lids. Be aware of the new legislation on feeding equipment, as above.

Beware of older products

If you buy or are given second-hand feeding equipment, it will not conform to these safety standards if it was made before January 2007. To be secure in your choice, always buy from an approved retailer such as a baby shop or pharmacy. Avoid cheap outlets like £1 stores, where the goods may be too old to conform to the standards, or may have been imported from a country that does not follow these guidelines. Check the safety numbers on the packs and choose products from companies that are members of the Baby Products Association, the trade association of the UK's nursery manufacturers.[4] Teats should be inspected regularly to check for wear and tear.

> ECO TIP: For a piece of furniture with a longer life span, look for high chairs that break into two parts to make a junior chair and table when baby is around two years old. Kids love having their own, baby-sized furniture.

High chairs

A really useful piece of equipment, your baby will eat, play and even nap in his high chair from when you start to wean until he is around two years old – or even older! There are lots of different styles, made out of different materials, though they broadly fall into two categories – wooden, and those made from a mixture of metal and plastic. Always ensure that the child is strapped safely into a high chair – 5 point harnesses are the best.

Chairs that grow with baby

These are the best option environmentally speaking, especially those made from FSC- or PEFC-certified wood. There is a growing trend for high chairs that turn into chairs suitable for a toddler, a child and then even an adult, using accessories to adapt to the age of the user and providing the correct seat height as well as good foot and leg support.

Whilst there are certain advantages to the plastic and metal high chairs available, such as having a reclining seat and adjustable seat heights, they do not score as high on environmental grounds. If you do decide on a metal and plastic version, try to get a second-hand one and (as always), pass it on after use. Look for a chair with a wide base – it will be more stable.

Lunch on the go

If you are often out for lunch, at cafés or at friends' houses, a portable high chair may be useful. Green options include a fabric harness that attaches to a standard dining-chair and a little wooden seat which fits on a chair.

Warming bottles

Electric bottle-warmers can be used to warm your baby's milk to the right temperature. However, you can avoid using the excess energy and resources, by simply placing the bottle in hot water; always check the temperature of the milk on the inside of your arm before feeding. Of course, if you can, the best way to be sure the milk is warm enough is to breast-feed, then the milk is always supplied at the right temperature.

Sterilisers

It is recommended that you sterilise all feeding equipment until your baby is one year old; this guards against bacterial infections, which can cause a baby to become ill and dehydrated very quickly. There are three ways to sterilise baby's bottles:

* *Boiling*

   The best method, as it uses no chemicals and no extra equipment which has to be manufactured and transported. Just boil all equipment, which must be fully submerged, for ten minutes in a large pan.

- *Steam sterilising*

  Whilst these electric plug-in versions can take from as little as eight minutes to warm bottles, they use energy, are an extra piece of equipment which has to be manufactured and transported and are usually made from non-environmentally friendly plastic.

- *With sterilising tablets or fluid*

  With this method, you place sterilising fluid or a tablet in a large container with water and all bottles and feeding equipment are immersed in it. It is a very mild bleach and not toxic at the levels used. However, if you are concerned about chemicals in your baby's feeding utensils or in the environment, use the boiling method. Bear in mind that the chemicals must be produced packaged and transported, and you will need to dispose of the solution after use, which all creates environmental pollution.

> ECO TIP: If you do buy a steriliser unit, ensure it is one that fits the largest quantity of bottles inside it and sterilises enough for the whole day at once. Never use it half empty, or you'll be wasting energy.

SLEEPY TIME

Bedding

Safety

Babies should not be put in a cot with a pillow, a duvet or quilt, or a fleece blanket until they are at least one year old; pillows can suffocate, and duvets and fleece can cause the baby to overheat, a factor which is implicated in cot death.

The baby should always be in the 'feet to foot' position – that is, the baby's feet touch the bottom of the cot and the sheets and blankets are folded in such a way that they come to the baby's shoulders and no higher. That way, he can't wriggle underneath the covers and get too hot.[5]

What do you need?

As a starter set, two fitted or flat bottom sheets, two top sheets and two cellular blankets are enough. You may want to add to this as you go along to save washing so often and to have spares or different colours.

It is fairly easy to find well-priced, good quality bedding made from natural fibres and even organic fabrics, made without chemicals and bleaches and there are now many outlets offering organic cotton and wool bedding (see *Resources*, 4).

ECO TIP: Ask around amongst friends to see if they have used organic cotton sheets or woollen blankets which they can pass on to you – and when you have used the bedding, you can pass it on to a friend, thus extending the lifespan of the bedding and using fewer resources.

Avoid non-organic fabrics

It may seem natural and instinctive to avoid fabrics that have been treated with chemical dips or sprays but there are also sound scientific reasons why you should do so. Pesticides get into our food chain or products, such as fabrics made from sprayed crops, and can affect people eating or using them; many pesticides are thought to cause the rise in the incidence of health problems such as birth defects, reproductive disorders and breast cancer.[6]

Pesticides also create health problems for those who work with them – see the Women's Environmental Network (WEN) for more information.[7] For more details on modern farming techniques and the chemicals used, plus the benefits to your baby and the planet of choosing organic fabrics, see page 71.

It's not just crops that can cause concern. Sheep in this country are routinely dipped in a bath containing organophosphates which can cause symptoms similar to M.E. or Chronic Fatigue Syndrome, according to a study by The Institute of Occupational Medicine in Edinburgh.[8] In addition, the wool is treated and bleached before dyeing with chemicals like peroxide.

It makes sense to put your baby in bedding made from fabrics that are as natural as possible and will not adversely affect his or her health.

Man-made fabrics

Try to avoid man-made fabrics used for sheets and blankets, like nylon, polyester and polyester-cotton mixes. Not only are they artificially produced but some are also quite sweaty fabrics, which don't allow your baby's body heat to escape efficiently when they are overheated. This is also true of blankets made from the artificial fabric called 'fleece' (which is actually 100% polyester) – used extensively for clothing like jumpers and jackets and increasingly for children's blankets as it is warm, soft and can easily take bright dyes and patterns. Don't be fooled into thinking this is some sort of natural fabric because of the name.

Changing units

Changing units are a place to store all the bed and bath-time stuff needed for cleaning and changing your baby and are positioned at a good height to prevent back-ache – but do you need one? Have you got a piece of furniture than can be used for the purpose until you no longer need it?

If you have to have a changing unit, try to choose one made from FSC- or PEFC-certified wood; you'll have to investigate carefully, as many units have MDF or plywood carcasses topped with wood veneers. Ideally choose one that turns into a piece of furniture later on, such as a chest of drawers, which will be extremely useful in the nursery and will last for years. Some suppliers stock rubberwood changing tables, a sustainable wood, though be aware that some models do contain some MDF.

> ECO TIP: Use a changing mat or a folded towel on top of a chest or chest of drawers, on your bed, or on top of a blanket box, at which you can kneel. Never leave a baby on a changer unattended – even very young babies are capable of surprising amounts of movement, so the lower, the safer.

Avoid purpose-made changing tables which have no other use

Changing units which are made with tubular metal legs and a plastic mat on top can generally only be used as changing tables and have no further use – and they are not degradable or recyclable.

Cots

One of the most important pieces of equipment you will buy for your baby is the cot, as he or she will spend a lot of time there, sleeping, napping and playing.

There are many styles, shapes and colours available but to ensure a cot is environmentally friendly, ideally look for one made from wood, preferably wood that is from a sustainable source in a managed forest and that is FSC- or PEFC-approved. For more information contact the manufacturer directly – most have excellent customer services departments and several have ethical statements and can tell you where they find their materials and how their products are made.

A bed for years

You can save money and resources in the long run by paying a little more at first and buying something that has a longer lifespan than the birth-to-two-years cots you can buy. There are a couple of options:

- Cot Beds: These are assembled in the first instance with rails on either side, looking very much like a traditional cot. Later on, when the child is old enough for a bed, the sides are removed and often the head- and/or foot-board splits in two so that you now have a neat little bed, suitable for most children up to around eight years.

- A cot that morphs into another piece of furniture, e.g. into a toddler bed, sofa, or even a desk.

- Both these options mean that you will be using the piece of furniture for longer, thus using fewer raw materials, less energy, and saving on transport costs.

ECO TIP: Make sure you keep all the parts and instructions safely so that when your child is ready for a larger bed, you can reuse it for a younger child or pass it on to someone with a new baby.

Possible hazards

There are a couple of potentially problematic materials used in cots, especially in some older models:

Lead paint

A second-hand cot may seem like a very green choice, but do check out how old it is. If you have fallen for an antique cot or one which was made before 1988 – perhaps the very cot you slept in as a baby – it may contain lead in the paint. If you use this for your new baby, he could be exposed to lead poisoning, particularly if the paint is cracked or peeling. You can even be at risk of harming your baby during pregnancy if you handle it or attempt to re-finish it when pregnant.

Symptoms of lead poisoning include being tired and irritable all the time, loss of appetite, stomach pain, poor attention span, bad sleeping patterns and constipation; lead poisoning can cause long-term or even permanent health damage.

If in doubt, either reject the cot (even at the risk of upsetting a doting grandparent) or have it stripped and repainted. It's not a good idea to do this yourself as improper removal of lead-based paint can cause you to be exposed to its effects. See the Government's notes on the safe removal of lead-based paint on the DEFRA website.[9] See page 14 for tips and advice on buying second-hand baby products.

MDF (Medium Density Fibreboard)

As mentioned above, this composite has been linked to health hazards and is best avoided.

If you are in any doubt as to the materials in your cot, avoid using it altogether.

SAFETY CHECK

Always check a cot to ensure it complies with current safety laws; look for the numbers BS EN 716 and BS EN 716-1:1996 on the label.[10]

Mattresses

The eco mattress – a natural bedtime

If you want your baby to sleep in and among natural, organic fibres that have not been chemically treated and have not been exposed to pesticides, bleaches etc., there are some wonderful, natural mattresses available. This is an important consideration if you are concerned about skin allergies, eczema, asthma and other health problems.

When you are buying your cot, cot bed or Moses basket, either ask the manufacturer if they offer an organic mattress (which may cost extra) or check that you can buy it without a mattress. Remember that cots and cot beds come in all different shapes and sizes, so ensure you have the dimensions to hand when you order.

Organic mattresses are made with fibres created by traditional farming methods, without the use of toxic fertilisers, pesticides, herbicides and fungicides. This is not just good news for farmers, wildlife and the countryside it also makes the products safer for your baby too. As we spend up to a third of our lives in bed (and babies much more than that), it makes sense to choose products that are free of chemical contamination. There are several UK companies now offering natural or organic mattresses, many of whom are able to offer sizes suitable for cots, cribs and Moses baskets (see *Resources*, 5).

There are several materials that can be used in the construction of an organic mattress:

Coir, coco mat or coconut fibre

This material is made from organic coconut fibres, taken from the outer husk of the nut. It is rather pricey as, at present, there is only one organic coir plantation in the world! It forms a supportive and breathable fibrous layer that is natural, has no chemical or artificial additives and is long-lasting.

Cotton cover

Organic mattresses usually have an unbleached cotton cover, fabric which has been grown without pesticides and which has not been chemically treated.

Lambswool or sheepswool

Wool is naturally resistant to harbouring dust mites and acts as an excellent thermal insulator, ensuring the mattress is cosy in winter and cool in summer. It also has natural flame-retardant properties, avoiding the need for chemical fire-retardants which can leach damaging fumes.

Mohair

Shorn from the Angora goat, mohair is a fibre that has been used for centuries and is not only silky and soft but also durable and resilient. It is also very good at regulating the temperature in the cot, being warm in winter and cool in summer.

Natural latex

Latex is a natural, springy substance made by tapping the rubber tree and collecting the sap. It is often used in mattresses, as it is comfortable and soft, yet firm. It will typically be covered with a layer of wool, cotton or other fabrics. Although latex can be responsible for skin allergies,[11] it is used in small amounts in an organic mattress and is well covered, so should not cause a reaction even for someone with an allergy. If in doubt, ask the company to send you a sample of the materials they use in their mattress.

ALLERGY ADVICE

If your child suffers from any allergy, cover the mattress with a good cover that provides a barrier to house dust mites and other allergens. Get one made from unbleached, undyed, 100% cotton with a very close weave that prevents the mites from crawling through them (see *Resources*, 6).

Conventional mattresses

If you have a history of allergies in the family or are simply concerned about having your baby sleep on a product which has been chemically treated, this information should help you make an informed choice when buying your baby's mattress.

Fire-retardant chemicals (polybrominated diphenyl ethers, or PBDEs)
Used to reduce the flammability of a mattress, these may interrupt brain development, interfere with thyroid hormone levels and may cause cancer in mice. Animal studies also reveal that there's no known level at which these health effects do not occur. PBDEs can cross the placenta and have been found in breast milk. [12]

Latex
Used for padding, this has been shown to cause allergic skin reactions and possibly asthma. [13]

PVC (Polyvinyl chloride)
Used to seal the outside of the mattress as a cover, PVC is thought to cause cancer, birth defects, genetic changes, chronic bronchitis, ulcers, skin diseases, deafness, vision failure, indigestion, and liver dysfunction. May give off fumes and can also be incredibly hot to sleep on, especially in the summer. [14]

Polyurethane foam
Used for padding, it can harbour bacteria and can cause bronchitis, and skin and eye problems. It may also release toluene diisocyanate, which can produce severe lung problems. [15]

SAFETY CHECK

A cot mattress should comply with British Safety Standards BS 1877 and BS 7177. It should be smooth, firm and fit the cot with no more than 4cm anywhere between the edge of the mattress and the edge of the cot. If the gap is bigger the baby could become dangerously trapped. The recommendation for bar spacing in a cot is between 2.5 and 6.5cm to stop babies getting their heads stuck, or worse. See the guidelines from the Baby Products Association. [16]

Avoid second-hand mattresses

It is recommended that each new baby has its own new mattress. There have been reports that old mattresses may harbour germs and bacteria, particularly *Staphylococcus aureus (S. aureus)*, and this in turn has been linked to the possible causes of Sudden Infant Death Syndrome (Sids) or cot death. In any event, a previous baby will have vomited – and worse – on the mattress, so it's certainly more hygienic to buy a new one!

The Foundation for the Study of Infant Deaths updated its advice on cot mattresses in September 2005 in light of recent research. [17]

ADVICE FROM THE FOUNDATION FOR THE STUDY OF INFANT DEATHS (FSID)

- Keep the mattress clean and dry

- Ideally you should buy a new mattress for each new baby. If you can't, use the one you have, as long as it was made with a completely water-proof cover and has no tears, cracks or holes

- Clean it thoroughly and dry it

Check that the mattress:

- is in good condition

- doesn't sag

- is firm, not soft

- fits the cot without any gaps

Ventilated mattresses (with holes) are not recommended as it is not possible to keep the inside clean.

Monitors (baby-listening devices)

There are many types of monitor, so you need to do your homework before buying as there are numerous portable wireless units for sale that transmit using electromagnetic radiation. Digital monitors, due to the 'pulsed' nature of the signal sent, are considered by some to be especially harmful to babies. Buy a monitor that either transmits the signals via the household ring main (they are simply plugged into existing plug sockets) or uses wires.

Do you want to reassure your baby without going into the room (great when you're trying to teach them to get to sleep on their own)? Then a unit through which the baby can hear your voice is great.

Monitors work either on battery power or from the mains. If you choose a battery version, invest in a battery-charger and a set of rechargeable batteries – quite apart from being green, this will save you a fortune in the long run and cut out buying packet after packet of AA batteries. They will be pressed into service later on for electronic toys with silly beeps. Ah, the joy!

Make sure you dispose of dead batteries safely – ring your local council for further advice.

Moses baskets or cribs

Both these items are used for only a very short time – as little as three months in the case of most babies. However, many of us want to put our babies to sleep in something small and cosy when they are tiny, as they look and feel so much more comfortable than in a cot or cot bed, which looks huge at first.

As with other nursery furniture, ask around for a slightly used one from a friend or family member (and it will only be slightly used). Some families have a basket that is used for each new baby as it comes along – so if yours does not, start a new tradition. Remember to give a second-hand crib or basket a good clean with mild soap and water and buy a new organic mattress and organic bedding.

If you buy one, most baskets are made from wicker, which is biodegradable and is from a really sustainable source, so it makes a good choice. Otherwise, look for a wooden crib and either use it for more than one baby or pass it on to someone else (see Chapter 7).

Travel cots

A handy addition to your nursery if you do lots of travelling, particularly if you are visiting relatives or friends who don't have children. But do you really need one? Here are some alternative ideas:

If you are visiting a hotel or guest house, or even a hired holiday home, the owners may provide a cot or travel cot. Check before you go. Bear in mind that this mattress will not be new and will have had other babies sleeping in it, so if you are happy to use it, take along your own bedding so you can be sure of the fabric next to your baby's skin.

Some local baby shops hire out travel cots and some also offer three-wheel buggies and rucksack-style baby-carriers for holidays in the country. If your local store doesn't – suggest it. They make the money back on the rental fees quite quickly, so they may welcome the idea.

Another smart option is a travel and play 'bubble'. It only weighs around 3kg and as it is suitable for sleep and play, it will get more use – and though it does use some man-made materials in its construction, it does have a cotton mattress and, being so small, uses far less raw materials than some of the full-sized travel cots.

THE GREAT OUTDOORS

Car seats

An essential item in our car-led society, the law states that your baby must always be secured into a car seat suitable to his/her age and weight, even for the first trip home from the hospital.

The issue of which seat to buy needs a lot of thought and planning, as you will want to get the best car seat for your needs without spending a fortune – and without getting it wrong! Have a chat with the owner or manager of your local independent baby shop or in one of the larger stores. Many members of staff will have been on a training course and there should always be someone to answer your queries. If not – look elsewhere.

The Baby Products Association (BPA) has an advice leaflet on buying a car seat.[18]

So, with very few exceptions, you know that you need one – but how can you ensure that your choice makes best use of resources?

Second-hand car seats

- It is recommended, for safety reasons, never to buy a second-hand baby's car seat. If it is involved in a crash, or dropped, it may develop stress fractures in the plastic and be unsafe in the event of an accident.

- Beware of buying from a second-hand shop or an online re-seller, as you may not get all the correct straps or fitting instructions, so the seat may not be correctly fitted.

- Although this is all very important for a baby's safety, the downside is that car seats with absolutely nothing wrong with them are put on the scrap-heap after a year, adding to the piles of non-degradable products in landfill sites.

- However, you may have a friend or relative who can pass one to you and help you fit it, which will be a good use of resources. Be sure you are confident that the person would tell you if it had ever been in an accident.

CAR SEAT STAGES – A QUICK GUIDE

Group	Weight	Approximate Age
Group 0	birth to 9kg	birth to 9 months
Group 0+	birth to 18kg	birth to 1 year
Group 1	9 – 18 kg	9 months to 4 years
Groups 2 & 3	18 – 36kg	4 to 11 years

A QUICK GUIDE TO PUSH-CHAIRS

Push-chair

A generic term used for many kinds of baby transport.

Three-in-one pram

This typically has a carrycot (a cosy bed in which the baby can lie flat, some-times with a sit-up base too), suitable up to six months, a car seat that attach-es to the chassis and a push-chair seat, usually suitable from six months upwards, though some from birth.

Two in one pram

As above, but with no carrycot. The seat reclines for a newborn. Check that it goes completely flat.

Travel system

A push-chair or pram that comes with a first-stage car seat, that can be swapped easily from being clipped onto the push-chair to the car. Ideal if you do lots of short trips in the car and then use the push-chair at times when baby is sleeping. Be aware, though, that the best position for a baby to sleep is flat on his back, not in a car seat.

Buggy or stroller

Usually suitable from around four or six months, these are small, lightweight, usually umbrella-folding push-chairs which are useful for short trips and for easy storage. Great for putting in the boot of the car, using on public transport and for holidays.

Three-wheeler or all-terrain pram (ATP)

The push-chair of choice for parents who do a lot of walking on rough ground, or even for those who want to jog (they were initially designed for this). The large wheels help the push-chair glide over bumps. They have become some-thing of a fashion statement in recent years.

Avoid buying unnecessary car seats

The law states that all children under 11 years or under 135 cms (4' 5") must be seated on an appropriate car restraint.[19]

Parents will often buy

- a Group 0+ seat (suitable from birth to around one year)

- a Group 1 (suitable from nine months to four years)

- a combination seat which turns into a booster, suitable up to 11 years.

You can reduce your number of car seats to two by:

- Getting a seat that is suitable to use from birth and lasts for around four years (Group 0+ and 1) and then buying a second seat which lasts from four to eleven years (Group 2 and 3)

- Cutting out intermediate stages by purchasing a Group 0+ (birth to 12 months) and then one that lasts from nine months to 11 years (Group 1, 2 and 3).

To keep up to date with the latest in the laws on car seat fitting and for help and advice on which category of seat is best suited to both your child and your car, look at the Government website, as above.[20] You can also get information from the car seat manufacturers, who provide guides on which of their car seats are correct for the weight and height of your child.

> ECO TIP: Get a second-hand push-chair – you may find just what you want, barely used and with all the accessories, at a fraction of the price. And remember, it's not second-hand – it's 'pre-owned', 'formerly loved' or 'recycled'!

Changing bags

It is useful to have all the things you need for your baby handy in one place when you go out – nappies, change of clothes, food, wipes etc.

Rather than buy a bag designed for that purpose, why not just adapt a bag you already have and save the money. Slip in a folding changing mat and you can use a favourite bag instead of one that you will not be able to reuse when you no longer need it and that will only make you think of dirty nappies!

Prams, push-chairs and buggies

There are several names for the wheeled vehicle you put your baby in (see below) and you may already have a good idea of the kind of push-chair you want. Many new mums set their hearts on a particular push-chair, perhaps for the fabric, perhaps because it has a car seat that clips on to it, or perhaps because it has a carrycot. However, all too often, you find that the item you bought is impractical, too fiddly or simply doesn't fit in with your lifestyle or fit in your car, and, if you get it wrong, you may end up with two or even three push-chairs – which is a huge waste of resources.

Push-chairs are all made of metal, plastic, rubber and chemically treated fabrics, they are mostly not biodegradable and will have used a lot of energy, raw materials and labour to produce, so being 'green' doesn't seem possible at first glance. However, there are things you can do:

- Choose right the first time – which means that you will buy just the one push-chair

- Pass it on when you have finished with it, which will extend its life

- Or get a second-hand push-chair

Make a good decision

- Ignore things like fashion, colours and fabrics (the pram industry is geared up to selling more and more products, so fabrics are changed every year – even though the basic pram underneath is still the same. I once heard a new mum say, about a pram that was only about four months old, "Oh, that's the OLD model.")

- Think about your lifestyle and how you will fit the baby into your family and routine. If you use public transport a lot, a small, lightweight buggy, suitable from six months, would be best.

- Why not use a sling or carrier until baby is old enough to use it?

- If the boot of your car is small, an umbrella-folding push-chair is a must.

- If you live in the country and do lots of walking, especially in woods or fields, then a three-wheeler is a good choice, or consider a back-pack.

- Ask yourself if you really need a carrycot – it can only be used for a short time. If you have a Moses basket, it may be unnecessary – use the carrycot instead of a basket or crib.

- A push-chair that can take you from birth to three years is a great investment. It will generally have a seat that can lie flat for newborns, and will have two or three sitting-up positions. This means you can go out shopping, or to friends and always have somewhere for baby to nap.

- Washable covers keep the push-chair looking good and extend its lifespan. This is particularly useful once your baby starts eating finger foods like rusks, rice cakes and croissants, which have to be the messiest foods on earth! Wash with an environmentally friendly laundry liquid or powder and give it the occasional 60° wash to kill bacteria.

- Check the build quality. If you buy a sturdy pram, it will not only transform into a buggy as the child grows but will also be able to be used for more than one child – then, when your family is finished, pass it to a friend.

Push-chair accessories

If you buy a new push-chair you will be offered all sorts of accessories to go with it – cosy-toes, rain covers, sunshades etc. The best way forward is to resist buying it all when you are getting the push-chair, as there are some things you may not need – and some you may not even use.

Sunshades

The push-chair's hood often does good duty as a sun shield.

Cosy-toes

A blanket from the cot can double as a cosy-toes.

Rain-covers

A good idea in this country: buy a strong, well-fitting one (the manufacturer's is usually best as it fits properly and will be well made), rather than a cheaper flimsy version that may rip. This is another example of where a higher outlay initially will not only save you money but save resources too, as a stronger, better quality cover will last and last.

Slings and baby-carriers

These are great and mostly tick all the boxes on the environmental front. They tend to be made of good quality cotton or linen, they are washable, adjustable and reusable.

Borrow one from a friend or buy one second-hand (many are used for only a short period of time but be sure to check for wear and tear to the fabric). Try out the different styles before buying to see which suits you best, whether it's a front carrier or a sling in which you can carry a tiny baby lying across your front, or a toddler supported on your hip. You can also get rucksack-style versions to go on your back, suitable from when your baby can sit, which are excellent for holidays and travelling.

Safety equipment

This is essential in modern homes and will save you many an accident, trapped finger and tumble on the stairs. With many items, such as socket covers, drawer and cupboard locks and bath mats there is little or no alternative to plastic. So gain your green points with other items such as using wooden stair-gates.

Bath mats

A very useful bath-time aid, this stops baby slipping about too much and also makes the bottom of the bath softer and warmer for tiny bottoms. If you bath with your baby, this will also be an important safety item for you – make sure someone passes the baby in and out of the bath to you. You will find lots of these in second-hand shops and though at first they may look grubby, due to the mould that these items attract, a quick scrub and clean up will make it serviceable for many years. Remember to keep the mould at bay yourself and clean regularly, to avoid the build up of germs, and you can then pass it on to a friend – or continue to use it for your toddler.

Drawer and door locks

Though you might think you need these on every drawer or door from the day of your baby's birth, you can minimise the amount you buy by waiting to see what the problem areas are. This won't be until your baby is crawling at least, and then you might be surprised at what you don't need – the baby may tend to be in one particular room most of the time, so you may not have to put locks in other places.

They are essential on cupboards that contain potentially dangerous substances, such as cleaning products, medicines and other hazardous items. Move these into high cupboards, where babies and toddlers can't reach them. If you do buy locks, make sure they are screw-on rather than stick-on, as the latter tend to come off and can't be reused. Keep all the instructions and spare screws etc. all together in one place and take them off when the need for them is past – and recycle them!

Plug covers

It is really easy to get these second-hand or from friends who no longer need them. Put them in all sockets within reach of a crawling or toddling child.

Stair-gates

The most environmentally friendly choice is wood, provided it is from a sustainable source; and it has the advantage of blending well with furniture. You may be able to pick one up second-hand.

Alternatively, if you choose a really durable metal gate, this will last through several children and still be rugged enough to pass on to a new family.

As with cot beds, make sure you keep all fittings, screws, allen keys etc. with the instructions when you are not using it and always check it has all the right pieces if you buy one second-hand.

Chapter Two

Gifts and Toys

When a baby is born, it is natural for the parents, relatives and friends to want to mark the occasion with gifts for the new child. This may be in the form of presents given to the baby when he or she is born, or it may be at a christening, naming ceremony or party shortly after the birth. A 'Baby Shower', a fairly established celebration in the US but one that is becoming increasingly popular here, is another way for people to celebrate the event.

As parents, you will often be asked what you would like for your baby, and this can be a great opportunity to mark your baby's birth in keeping with your personal ethics, wishes and preferences. Here are some ideas of green and ethical presents which you – and your baby – will love!

A BRIEF HISTORY OF BABY GIFTS

Why do we give gifts to a new baby?

As with many of our traditional feast days and ceremonies, a baby's naming was, historically, a celebration. It marked a way for the child's family to give thanks for the safe delivery of a healthy child and it officially welcomed the child into society.

The practice of giving gifts to a new baby or its parents goes back many centuries and is found in many cultures. It celebrated the arrival of a new member of the community and provided the opportunity for other symbolic blessings – for example, the giving of precious metals or costly gifts to the child was intended to symbolise wealth and prosperity in later life.

The christening of the first Queen Elizabeth in 1533 is a good example of how expensive gifts were lavished on a child who could not possibly appreciate them – she was given cups of silver gilt and gold and was dressed in a mantle of purple velvet trimmed with ermine! More recently, Elizabeth Bowes-Lyon, later Queen Elizabeth the Queen Mother, was given a brooch in the shape of a bumblebee, set with diamonds and gems. Silver, gold and jewels, it seems, have always been popular gifts.

What is traditionally given?

In the UK, certain gifts have become traditionally accepted for a new baby: for girls, these tend to be tiny, baby-sized bracelets or necklaces and lockets, usually made from silver or gold; and boys, paradoxically, are given a beer tankard in silver or pewter. These have always struck me as slightly odd gifts – the girls get a present they are unable to wear once they grow a bit too big for the jewellery and the boys have to wait 18 years for their first beer in the tankard! The idea behind these gifts made of precious metals is to offer the child prosperity and a wish for their future life to be one of ease and plenty.

Rattles and soothers are also traditional gifts, ones which go back centuries, sometimes combined into one item to both amuse the baby and help with sore gums while teething. The Romans used teethers made of peony wood beads that were not only satisfying for the baby to chew on but were supposed to have medicinal properties too. Coral was another popular material as it was supposed to ward off evil spirits and illness as well as being cool and soothing to the teething baby's mouth and satisfyingly knobbly. Other materials included bone, ivory and semi-precious stones like crystal and agate, often set in silver.

Christening and naming gifts available today

As our consumer society has grown and more and more products for every kind of occasion are produced, there is a much greater range of christening gifts available, from jewellers, card and gift shops, department stores and online.

Silver remains a popular material for gifts and items include not only jewellery but also keepsake boxes (which can be used to store your baby's first tooth or a lock of their hair), money boxes or piggy banks, mugs and rattles, photo frames and albums.

China is also popular, with products like a feeding set (bowl, cup and plate), or an egg-cup and spoon. Other items offered include a baby's first brush and comb, height charts, name banners and miscellaneous items like a christening certificate box or holder.

Personalising these items is big business. You can get all kinds of things personalised with printing, engraving or embroidery – from teddies to dolls, towels and even blankets.

These are some of the gifts which you may find you are offered when your baby is born. The following pages will give you some ideas and inspiration for environmentally sound, ethically produced gifts so that your baby's birth can be a truly eco-friendly event.

Eco-friendly gifts

If you are thinking about what presents you can ask for on behalf of you and your new child, there are plenty of 'green' options – and there are some things you may actively want to avoid, too, because of their effect on the environment or disadvantaged people. Don't be shy about asking for specific items and making your green preferences known – many people are only too pleased to be offered a little guidance.

Gifts to the planet

Trees

One lovely, personal and long-lasting way to mark the birth of your baby is to plant one or more trees to commemorate the event. This contributes to the environmental health of the planet as trees take carbon dioxide from the atmosphere and lock it up for their lifetime.

There are many options available:

- A tree for the garden: If you have a garden, you could plant your tree there – though ask yourself if you would be upset if you moved house and couldn't take the tree with you!

- A tree which can move house with you: If you want it to travel with you wherever you go, opt for a compact tree that you can plant in a large pot, such as a decorative magnolia or maple.

- Several trees: If you are lucky enough to own a piece of land, you could plant a number of trees or an orchard, which will bring pleasure to many for decades and provide an important habitat for wildlife. Try to ensure that the trees chosen complement local woodland.

- A tree for your area: Find out whether your local council has a tree-planting policy that you could contribute to. That way, your tree could be in a public place – a favourite park or green space or even along a road. For example, the London Borough of Redbridge has a scheme whereby it will fund half the cost of the purchase and planting of the tree and the person requesting it will fund the other half. Residents of Dartmoor National Park can apply for assistance in tree-planting from the National Park Authority. Call your local authority or one that has responsibility for nearby woodland to see what is available.

- A tree for a charity: Contribute to one of the organisations who plant trees around the country and who work to save areas of threatened woodland by purchasing land where the trees would otherwise be cut down. The best known of these is The Woodland Trust which campaigns for the protection of ancient woodland and trees across the UK. Ancient woodland, land that has been wooded for at least 400 years, is our richest habitat for wildlife and is irreplaceable. Ancient trees also have great cultural, historical and ecological significance. The Trust offers a range of gifts that can be used for commemorating special occasions such as the birth of a new child. You can dedicate the trees in a wood of your choice from the 20 areas they control around the country. You will receive a card, a window sticker, a certificate bearing your dedication and a map of your chosen wood showing where your tree is planted and other information about the woodland (see *Resources*, 1).

- A 'Family Tree': an oak sapling in a lined wicker basket (which you can use for other things), presented with a soft toy, a pair of socks and a 'baby asleep' sign (see *Resources*, 2). Something along these lines can of course be put together by your friends or relatives who can choose their own sapling and gifts.

Gifts to an environmental charity

There are many environmental and ethical charities who would benefit from a donation on behalf of your child who have a catalogue and online shop, for example The World Wildlife Fund. There are several charities whose main focus is the environment, for example Friends of the Earth and Greenpeace. For more ideas, try websites that have links to companies who use environmental or ethical values as their starting point, for example Ethical Junction, which lists companies who produce ecologically sound products, or A Lot Of Organics (see *Resources*, 3). To find accredited suppliers of ethically traded goods, contact the Fairtrade Foundation (see *Resources*, 4).

Gifts to other babies

Many new parents are acutely aware, as they bring their baby home to houses with heating, lighting and water, that there are many children born around the world who do not have access to what we view as basic amenities. So you, or a family member on your behalf, could consider sponsoring a child in those circumstances to help give it at least some of the advantages – healthcare, education and housing – that your child enjoys.

When sponsoring a child, you will usually be asked to make a fixed, regular monthly donation, usually by direct debit from your bank account. The money is used by the charity or organisation to provide what is most needed in the community where your chosen child lives. What is especially rewarding about this is that you are given regular updates on the child's progress and the achievements the charity is making in their community and you can exchange letters and gifts. As your own child grows, they can participate in the exchange of information, photos etc. with their 'brother' or 'sister'. This is designed to be a long-term commitment, although the donors can pull out if their financial circumstances change (see *Resources*, 5).

Gifts to a community

If your friend or relative prefers to make a one-off gesture rather than committing to an indefinite outgoing, they can give a donation to one of the many charities and non-governmental organisations (NGOs) that work for better conditions in third world countries. There are many charities which offer a wide variety of choices of items that can be donated to a community, geared towards providing what is most needed (see *Resources*, 6). For example, you can donate a toilet to areas where sanitation is a key problem, helping improve sanitation standards and reduce water-borne diseases like cholera and dysentery. Items for schools are popular too – you can give textbooks, desks and chairs or kit out a teacher. South American farmers rely on their livestock as a source of wool for selling, and you can give them the gift of an alpaca together with the food and other items it needs for its first year. You can send mango saplings that will grow into fruit-bearing trees and thus provide an income for African farmers, or you can give tools to help the workers help themselves.

Gifts to wildlife

From adopting an endangered animal breeding programme at a local zoo to donating money to a wildlife charity, many find this a fulfilling act of generosity. If you buy a sponsorship for your baby, the zoo or wildlife sanctuary often puts up a plaque or sign with the sponsor's name on it (see *Resources,* 7).

Green savings and investments

Every British child born after September 2002 is entitled to receive a voucher from the Government worth £250 to start their own savings policy, and if your family is eligible for full Child Tax Credit, you will get a further £250.[1] This can be put into one of the Government's Child Trust Funds, accounts for which are available through one of several companies the Government has chosen (see *Resources,* 8). You and your friends and family can then invest up to £1,200 per year on top of this until the child's 18th birthday, all tax-free.

Some friends and relatives, particularly grandparents, may like to offer money as their newest grandchild's present, which can be put in a savings scheme until the child is 18 – and old enough to decide how to spend it! This may be savings towards university, a first car or even for a deposit on a home, depending upon how generous the relatives are and how the investment performs!

You may want to avoid investments in contentious areas such as nuclear power, arms trading, oppressive regimes, tobacco, alcohol, testing of drugs and cosmetics on animals or those which exploit third world countries. It also makes sense to choose one with a good record in increasing the value of its investments!

There are several banks or financial institutions that offer good environmental or ethical investment packages, such as investment in environmental protection, pollution control, recycling, energy conservation and equal opportunities policies. Check with the investment providers to see which issues they have concentrated on to find one which best reflects your values (see *Resources,* 8 and 9).

You may also like to consult an Independent Financial Adviser who specialises in green investments, such as the Gaeia Partnership, Ethical Investors Group and Barchester Green Investors (see *Resources,* 10).

Jewellery

It's traditional, it's easy to find and it lasts for years – if not for decades. But jewellery raises many environmental and ethical issues and may not be a good choice for the planet or for the workers in less fortunate countries who work for low pay in bad conditions. However, you can choose to buy jewellery which will help the planet and make the lives of the producers better, too.

* *Second-hand*

   If you've set your heart on a gold locket for your beautiful new little girl, look for old gold – maybe an antique Victorian locket or necklace – in a second-hand or antique jewellery store. You should also be able to find vintage christening bracelets. Then your gift will be a piece of history.

- *Recycled*

  There are companies that offer jewellery made from recycled gold, such as Green Karat (see *Resources*, 11). If you can't find something you like, a jeweller may be able to make something up for you using old gold. Ask at your local jewellery shop or call the British Jewellers' Association (see *Resources*, 12) for a recommendation.

- *'Ethical' jewellery*

  Ethical jewellery is available in several high street outlets, and charity shops – Oxfam have a great range, for example. By 'ethical', the producer may mean that it is made from recycled materials, it may be made by previously disadvantaged people who are being paid a fair wage, or it may mean that the materials it is made from are produced in a way which does not pollute or exploit. Ethical and fair-trade jewellery is now available from a number of reputable internet suppliers, too. One of the best websites I have found is Silver Chilli, a company bringing fair-trade silver jewellery to the UK from Mexico (see *Resources*, 13).

Here is a heart-warming story from Jane Kellas, founder of Silver Chilli Jewellery, about how their profits are used, which is typical of the way many fair-trade companies return profits and benefits to the people they work with.

"Silver Chilli is a fair-trade company committed to social investment whilst bringing fabulous jewellery from Mexico.

Each year we invest part or all our profit in a social project chosen by our workers. Last year that choice fell to the women's group who make the beautiful hand-embroidered bags in which the jewellery is presented. They live in an isolated village 12 hours by bus from the city of Oaxaca – in the second poorest state in Mexico. The women are aged from 30 to 75 years. Only about five of them speak Spanish and of them only two or three can write it; most speak 'Mazateco', which is the language local to the area. They have traditional lifestyles, which makes it difficult to earn enough money to live well. When we asked them to tell us how they would like us to make their lives a little easier, we were very surprised when they asked us for spectacles! But, being 12 hours from the nearest city and given the high cost of glasses, acquiring a pair can be quite a challenge for the group. So we flew an optician and his equipment onto a plane and to the village. Several hours later, more than 15 women had eye tests and were looking forward to receiving their specs.

We wouldn't be able to spend our profits in this way, if our UK customers hadn't helped us to generate them."

For other fair-trade jewellery stockists, see *Resources*, 14.

Toys

When choosing toys for your baby, you will naturally want to ensure that they contain no toxic plastics or paints and that they are safe for a baby to play with and to put in his mouth. Legislation in this country and the EU is fairly stringent on toys and you are generally safe if the toy has the CE mark on the label (ensure it is suitable for the age of your child – many are unsuitable for children under three if they have small or removable parts).

Wooden toys

Wooden toys are great environmentally friendly gifts as long as the wood they are made from comes from sustainable forests (FSC or PEFC). Wooden toys score much higher on environmental grounds than their plastic equivalents, which are ecologically unsound to produce and then, when broken or discarded, spend the next hundred years – or more – polluting the soil in landfill sites. Wooden toys are not only good for our planet, they can also be repaired and repainted for subsequent children to enjoy and when they are finally too old or broken for play, they biodegrade safely.

Visually stimulating wooden toys like mobiles are a great thing to have in the nursery – they will help the baby's eye co-ordination skills and give him something bright to look at (see *Resources*, 15).

Organic toys

There are several companies who are making soft toys from organic cotton which will not cause any skin allergies or irritation to existing allergies, and which have been manufactured in an environmentally responsible way (see *Resources*, 16).

Fair-trade toys

Several UK companies have links with third world producers who not only make toys for our little ones but preserve ancient crafts and skills at the same time, for example traditional knitting skills in Kenya and Peru. There are several co-operatives where the workers make hand-knitted garments and a fair price is paid for each, ensuring the workers can re-invest in materials, machinery and, most importantly, their community.

Many fairly traded toys have other benefits too, such as being made from organic and natural fabrics and materials. The dyes used to give them their colours are usually non-polluting and safe for babies and children (see *Resources*, 17 and 18).

Gifts you might want to avoid

Beautiful but deadly – the ecological legacy of goldmining

You may have read reports stating that gold is now considered quite a contentious metal, environmentally speaking – but as it has been discovered and mined, worn as decoration and used for precious objects for centuries, why is it suddenly having such an image problem?

The reasons lie in the way gold is mined today. Worldwide reserves of gold have been steadily declining but demand has continued to grow, both for jewellery and for use in science and industry. What little is left is now extracted in a way that is causing catastrophic damage to the environment.

The process, put simply, means that huge areas of land are devastated to produce small amounts of gold, leaving behind wastelands of scrap rock polluted by the cyanide used to extract it. The consequences are acidic water levels (called Acid Mine Drainage). The most acidic water in the world (10,000 times more acidic than battery acid) is found in caves in California's Iron Mountain Mines where metals including gold have been mined [2] that are hazardous to animal and human life. Workers live and work in poor conditions near to the mines and their continual exposure to chemicals and dust can cause cancers, fibrosis and poisoning, raising human rights issues.

Waste rock powder from mining is built up into huge 'dams' designed to contain it, but there have been many well documented disasters when they fail, the most notorious of which was when a dam collapsed in Romania in 2000. A slurry of around 100,000 cubic metres containing cyanide, copper, lead and zinc poured into the Danube, killing all fish and damaging fish-eating creatures for 250 miles of the river as it passed through Hungary and Yugoslavia before emptying into the Black Sea. This may have health consequences for the people living in areas affected by this for decades.

Another issue is the use of cyanide. Around 200 tons of sodium cyanide is produced worldwide every year, a staggering 180 tons of which are used in the gold industry. [3] Though cyanide quickly disperses on exposure to sunlight, it is highly toxic and hazardous to workers, not to mention flora and fauna near the mine.

Foreign flowers

You may not have a choice over what kinds of flowers are sent to you when your baby is born – but if asked you have the opportunity to choose ethically and avoid flowers whose cultivation and transport cause environmental damage. Why not buy from an online, ethical company such as Wiggly Wigglers, where you can choose to send local, seasonal flowers (see *Resources*, 19)?

People employed to grow, pick and pack flowers in the Third World are often paid poor wages and have poor working conditions. On the other side of the equation, however, is the fact that, for many areas of these countries, flower production may be the only local viable employment option; if you do choose flowers from abroad, look for fair-trade flowers. If we start to demand more locally sourced and fairly traded flowers in our florists and supermarkets, this will have a knock-on effect on the flower industry.

> ECO TIP: Buy local – flowers are grown in the UK mainland as well as the Channel Islands, Cornwall and The Isles of Scilly and are available by mail order, or buy from a local florist or grocer and choose flowers that are in season rather than exotics.

Name a star

Whilst it may seem a tempting idea to name a celestial body after your new son or daughter, the companies that advertise as 'official' star-naming organisations are not genuine at all. The only body which can officially name stars is the International Astronomical Union (IAU).

There are many companies on the internet offering to name you a star and providing official-looking certificates. The fact is that most of the stars have boring numbers rather than names, as scientists would rather look for star 64487 than Poppy Smith from Brighton!

Chapter Three

Nappies

A BRIEF HISTORY OF NAPPIES

The problem of how to resolve a baby's toilet needs is as old as civilisation. For centuries, mothers have used various types of cloth wrap and plant materials to collect or soak up their children's waste. Native Americans used the inside of the milkweed plant as an absorbent liner inside their baby slings, whilst Eskimos used a variety of moss.

European mothers have tended to opt for cloth wraps and these materials and the way they were fastened onto the baby probably continued unchanged for many centuries. They were most usually made from linen and there are references to these kinds of nappies from as early as Shakespeare's works. In Britain and the Commonwealth they became known as the Nappy (from Napkin). In America, however, the name of a kind of cloth with a repeating pattern, diaper, from which they were made, became more commonly used.

By the mid-19th century, major changes started happening as the Industrial Revolution changed the way cloth was made and priced – which coincidentally had an impact not just on nappies but on the lives of women. Thanks to improved trade methods, transport and the invention of mechanisation, Europe and North America began to produce cotton fabric which was cheap to buy and easy to launder – and so it became cheaper and easier for women to keep their babies clean. The invention of the safety pin in the 1840s also helped the modern nappy to take its form.

This state of affairs might have continued with as few changes as in the previous centuries, had it not been for the development in the 1930s of a soft cellulose tissue made from wood pulp. This was discovered in Germany as a by-product of the wood and paper industry and its absorbent qualities were noticed. A couple of years later in Switzerland, a paper company manufactured a soft paper parcel that could be placed inside a baby's pants.[1] The age of the disposable nappy had begun.

With the Second World War came the exodus of many women from their traditional place as homemakers, as they took up war work and adopted the jobs of the men who had gone away to fight. One of the results of this was that they began to need and demand more labour-saving devices in the home and the 1950s saw a huge increase in products aimed at women who were now needing or choosing to combine careers and families – the vacuum cleaner, refrigerator and washing machine – and the disposable nappy was seen as part of that trend. The added benefits for the baby – hygiene, keeping the skin soft and dry, and convenience were also important considerations for mothers.

For the twenty years or so following the war, both manufacturers and mothers experimented with different forms of washable and disposable nappy, aiming for the most convenient and comfortable hybrid. Disposables became more absorbent, thinner, better shaped and, most importantly, cheaper.

So far, so good. The new nappies had fulfilled a need and were making women's lives easier and babies were reaping the benefits in terms of comfort and health. They were also still, to a large extent, biodegradable. But an invention – or rather, the adaptation of use of another industrial by-product – in the 1960s, had consequences which no-one could have really foreseen at the time.

Starting in the 1950s and 1960s, nappy manufacturers started to replace bulky cellulose fibre with super-absorbent polymers – mostly a chemical compound called sodium polyacrylate – which can hold many times its own weight in water.[2] They also began to use more plastics in the construction of a nappy and adhesives. So whilst for the mum and baby it seemed as if this was a marvellous invention, making nappies smaller, more absorbent and less likely to leak, the ecological effects were not at the time really considered to be important.

It is only with the benefit of almost forty years of hindsight and a change in the way we think about our world and the effect we are having on it that nappies have become the bad guy.

Take a look around and you will find disposable nappies everywhere – cluttering up the waste bins of public toilets, thrown on beaches, pitched out of car windows and falling at the side of the road, in our domestic rubbish and most worryingly, in our landfill waste disposal sites. I have friends who have travelled to the foothills of the Himalayas, others to ancient native sites in Canada, places that are remote, beautiful and are – or should be – unspoiled. But they have reported that there was a trail not only of toilet paper but also of discarded disposable nappies as far as the eye could see.

Whilst potty-training my daughter, I asked my parents for their memories of attempting the same process with my siblings and me – I'm one of four children born quite close together. Whilst my mother – a consummate organiser – remembered the soaking, washing, folding and pinning (plus commenting that she had us potty-trained and out of nappies as soon as was humanly possible to avoid all the work), my father had very different memories. Like most dads in the 1970s, he was out at work most of the day, often only returning home after we were asleep. He says he remembered that for four or five years, there were nappies everywhere – drying on the line or on heaters, in piles in the bedroom. Then suddenly, they were gone as the youngest child graduated to the potty. No doubt my Mum and Dad were aware of the growing commerce in disposable nappies but at the time they were simply too expensive for the average family to use on a day-to-day basis.

Today, however, it's a very different story. Disposable nappies have gone down in price to the point where it is considered the norm to buy them rather than seeing them as a luxury item, or for occasional use. You are now 'normal' if you use disposables and are often viewed as a bit of a hippy if you opt for washables. However, this tide is once again turning as people become aware of what is now one of the biggest topics of debate in recent years – the environmental cost of disposable nappies and of the origins, effect and lasting legacy of their ingredients.

WHAT'S IN A DISPOSABLE NAPPY – AND WHAT ARE THE POTENTIAL PROBLEMS?

Here are the details of what goes into one of our most popular brands of nappies in the UK[3] and their potential problems:

Adhesives

Glues used to hold various parts of the nappy together.

What are the implications?
May cause irritation.

Aloe gel

Used as a skin calmer, may help prevent nappy rash and keep the skin soft.

What are the implications?
No known problems, a natural product.

Artificial perfumes or fragrances

Used to give the nappy a pleasant fragrance – that is, before the baby fills it!

What are the implications?
May contain phthalates, a plastic product which prolongs the life of a scent. May cause liver, kidney and lung problems and damage the reproductive system of developing babies. May also contain parabens, used as a preservative, which can cross the skin and build up in the body. May mimic oestrogen which can cause cancer.

Breathable outer cover

The mock paper outer used to make the nappy appear soft and flexible.

What are the implications?
Uses glues from petrochemical sources which may cause skin irritation.

Coloured dyes

Used to print cheerful pictures and characters on the outer layer of the nappy.

What are the implications?
May cause irritation, even nappy rash in some children, according to a study in Denver,

Colorado in 2000, where skin-patch tests showed that nappy rash was caused by the dyes in the nappies and training pants worn by children.[4] A later study in 2005 seemed to substantiate this theory, when researchers confirmed that nappy rash which didn't respond to typical treatments was found to be a result of an allergy to the dyes on the nappy.[5]

Petrolatum

A semi-solid mixture of hydrocarbons obtained from petroleum, often used in medicinal ointments and for lubrication. In a jellylike substance (known as petroleum jelly), it is used as a barrier cream and for lubrication.

What are the implications?
Drying to the skin, petrochemical derivative.

Polyacrylate

Also called sodium polyacrylate. A by-product of the petro-chemical industry, this is a super-absorber used inside a nappy. It literally sucks moisture away from the inner lining of the nappy and then holds on to it.

What are the implications?
It may cause irritation to the baby's skin, as it is very drying. Thus may cause nappy rash. In the past it has been linked to toxic shock syndrome, and was banned from women's sanitary products years ago. It is extremely difficult to break down in the soil – it may take 100 years or more.

Polyester

A plastic fibre used inside the nappy.

What are the implications?
Flammable. Can cause respiratory problems and allergic reactions if minute fibres get into the respiratory tract.

Polyethylene

Another plastic, often used as a coating for tablets.

What are the implications?
No known cause for concern.

Polypropylene

A flexible plastic material.

What are the implications?
May cause respiratory problems if inhaled; for example if the nappy splits.

Rubber

A natural product from the sap of the rubber tree; chemically altered versions are also used.

What are the implications?
May cause skin allergies.

Stearyl alcohol

Used as a skin emollient.

What are the implications?

Can be very drying to the skin.

Wood cellulose fibre

A by-product of the wood and paper industry, this has absorbent qualities. It is a natural product and can be sourced from managed forests.

What are the implications?

To make it so white, the fibre is usually bleached with chlorine, the by-products of which can affect water and fish in nearby rivers and lakes. It also remains in small quantities in the nappy and can cause birth defects, miscarriage, cancer and other health problems. It is toxic in quite small amounts.

NAPPY FACTS

3,000 nappy changes

A baby gets through an average of ten nappies per day in the early weeks of its life, this figure reducing to around six a day later on – that's an astonishing 3,000 or more nappy changes by the time he reaches his first birthday! [6]

Three billion thrown away

Around three billion disposable nappies are thrown away every year in the UK! [7]

It takes a long time for a disposable nappy to decompose – some sources such as Greenpeace quote over 100 years but the fact is, that we simply don't know how long they will take to break down in the soil – and what effects they may have in the future.

160 Bin Bags

Each baby is capable of filling 160 black plastic bin bags with used nappies by the time it is potty-trained – that's a lot of black plastic!

Over £1,000 wasted

Disposables can cost you at least £1,000 – probably more, as many babies who use disposables potty-train later than babies who wear cloth nappies.

Tax-payers pay for disposal

All of us who pay taxes are paying for disposables to be – er – disposed of. Cost to the average person? Probably around 10p per nappy. [8]

Other issues with disposable nappies

Can disposables cause or exacerbate asthma?

In 1999 a study in Atlanta, Georgia by a private laboratory showed an increase in asthma in laboratory mice exposed to cheap disposable nappies. Whilst the results could not of course be tested on babies, it may explain why the incidence of asthma has increased so dramatically over the last few decades. The mice seemed to be affected by the chemicals being given off by the nappies – a control of a cotton nappy was used, which showed no marked effect. [9]

Can disposables cause infertility?

Some scientists think that the plastic content of a nappy leads to a higher temperature around the scrotum of baby boys, which in turn may lead to problems producing sperm and thus infertility. A study published in 2000, [10] used thermometers to measure the scrotal temperatures of baby boys. They were found to have higher temperatures while wearing disposable, plastic-covered nappies than they had wearing cloth nappies. Since the production of sperm is dependent on the temperature of the scrotum being lower than the body, this may explain one of the causes in the rise in male infertility. However, there are thought to be several other environmental causes which may be more significant.

Raw sewage in the environment

If nappies are flushed down the toilet, they inevitably block the sewer as you'll know if you – or one of your kids – have ever tried it! But the alternative to disposing of human waste in our sewerage system (which is after all designed to treat and cope with the waste) is that it is disposed of in bins and ultimately in landfill sites. They can then breed bacteria and viruses, an effect exacerbated by the fact that the plastic outer acts as a mini-greenhouse. As many as 100 viruses can live and breed in them for up to two weeks including the polio virus from the nappies of recently vaccinated babies, which is potentially dangerous for those who have not been vaccinated. These bacteria and viruses can then also be washed into our water tables and affect rivers and streams.

How do you dispose of a disposable nappy?

When you dispose of a nappy, how do you throw it away? Chances are, you'll have bought a pack of plastic bags called Nappy Sacks or similar, in which you place the nappy before throwing it into another bin bag, also made of plastic. Or maybe you'll have bought a Nappy Wrapper, a great big plastic modified bin, which wraps each nappy in plastic like so many sausages before, in turn, they are thrown in the bin liner. This is the worst aspect of disposables – not only do they make use of non-degradable plastic in their construction but they are, when used, generally wrapped in more plastic, be it bags or rubbish sacks, before being thrown away.

If you do nothing else, please look for and buy degradable nappy bags and degradable bin liners, available in some supermarkets as well as health food shops. Alternatively, go to a supplier like D2W at www.degradable.net. If you get a whole box of them, it will be even cheaper than buying by the roll.

Do disposable-nappy-wearing babies take longer to potty-train?

Because the moisture is taken away from the baby's skin, The baby never 'feels' wet. This can lead to a delay in potty-training and as a consequence a longer time wearing nappies.

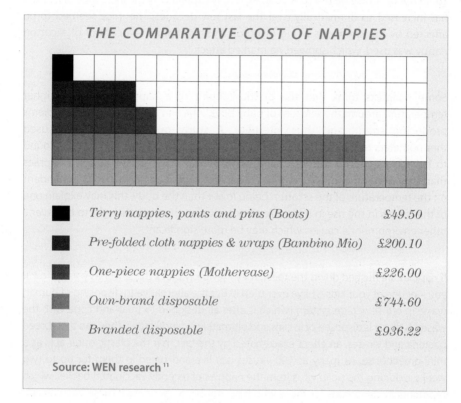

THE COMPARATIVE COST OF NAPPIES

▪	*Terry nappies, pants and pins (Boots)*	*£49.50*
▪	*Pre-folded cloth nappies & wraps (Bambino Mio)*	*£200.10*
▪	*One-piece nappies (Motherease)*	*£226.00*
▪	*Own-brand disposable*	*£744.60*
▪	*Branded disposable*	*£936.22*

Source: WEN research [11]

Washable nappies

I'm not suggesting that every new mum completely abandons disposables and goes back to the endless buckets, terry squares, pins and plastic pants that were used in the past. There are some great washable and disposable alternatives that can be used easily and efficiently and fit in with modern lifestyles. Don't set yourself impossible goals, then get disheartened if you fail to reach them. As a new parent you will be tired and stressed enough without giving yourself a hard time!

What I am suggesting is a manageable way of making less of an environmental impact and being sensible and flexible about it. It is still better for the planet if you mix and match – and you don't have to abandon your attempts to be green altogether if you happen to use a disposable once in a while! Remember, that every little bit you do now for the planet will mean that your child won't have to cope with the consequences when he grows up.

Better for the environment

Washable nappies are making a huge comeback in the UK, with more and more families choosing to use them. You may read reports that, when you add the cost of the washing, the amount of energy used and the detergents used to wash them, they cause as many problems as disposables. However, the studies that suggest there is no green advantage to cloth nappies are often biased and do not take all factors into account. For example, you can't put a price on these facts:

Cloth nappies are a reusable resource. Once they have looked after the nappy needs of one baby, they can be reused on two, three, four – in fact, as many babies as they manage before they fray or tear.

We don't know how long it will take disposables nappies to degrade in the soil. Conservative estimates put it at 100 years, some say 500. That's a terrible legacy to expect our kids and grandchildren to deal with just so that we can have a more convenient lifestyle.

If you just have these two facts at your fingertips, you will already be halfway down the route of being a cheerleader for cloth nappies. Be prepared for being viewed as an eco crank by some, and arm yourself with as many facts and figures as you can. If you need any more facts and figures to help with your decisions, look at the research conducted by WEN. [12]

Better for your budget

To put your baby in disposables can cost up to £936 at the time of writing, whilst washables, depending on the type and the number you buy, can cost from as little as £49 – and remember, they can be reused on several babies, making them even cheaper again – see the chart on page 51. In addition, many local councils offer their residents a cash incentive to use washables. Check with yours to see how much you can claim. To find real nappy sales, help and advice in your area, contact the Women's Environmental Network Nappy Line, set up in association with the Real Nappy Association (see *Resources*, 1).

There are three basic versions of the washable nappy:

1. Terry squares

Unfolded, simple squares of terry towelling. They need to be folded and pinned around the baby, then a waterproof pant used outside it. It is recommended that you use a liner, too.

Great for

Drying really quickly, they can be used for a boy or a girl (see folding tips), suitable for any size baby, very cheap to buy.

Bad points

Can be fiddly if you're in a hurry, waterproof pants can be hot for the baby to wear; they are not as fitted as some cloth nappies. Need pins or other fasteners.

How to use

There are tricks to the way you fold terry nappies: for a girl, you generally want most of the bulk of the nappy at the back, as urine tends to drain down towards the bum if the baby is lying on her back; for a boy, you want most of the bulk at the front as the penis usually points upwards (oh, and a top tip for parents of a new baby boy – they just love to pee just as you've taken the nappy off, and are capable of great feats of direction and range! Keep the nappy over the parts until the next one is ready to go on!).

First – the simplest fold. Place the nappy square on a flat surface. Fold it in three length ways – so, in from one side and then the other. Your nappy is now three layers thick. Then fold one of the thin ends about a third of the way in. This makes it six layers thick at that end. For a boy, place the thick bit at the front, a girl needs it at the back. Place a liner on the area that will be in contact with baby's bottom.

To put this on your baby, pick both the baby's feet up with one hand and slide the clean nappy under the bottom, making sure the edge comes up to the waist. Flare the back of the nappy out slightly to make flaps. Then gently bring them to the front and pin. Make sure you use safety pins (and put your hand between baby and the pin so you don't accidentally prick her), or use Nappy Nippas or Nappy Snaps (see *accessories*, page 58). Cover the nappy with a pair of plastic pants or a waterproof outer.

There are several other ways of folding terries to cope with different circumstances – for example, small-birth-weight babies, ones whose stools are very wet etc. Have a look at the many folding tips on the Nappy Lady website (see *Resources*, 2) where there are folds called things like the boy fold, the poo catcher and the butterfly fold.

Where to buy

There are a few companies who offer organic cotton nappy squares (see *Resources*, 3); try to find organic cotton versions if you can, as they are even more environmentally friendly than their non-organic counterparts. However, they can be expensive, so if you can't afford organic, look for cotton squares available at most independent nursery shops – if they don't keep them in stock, they can usually order them for you from their wholesalers.

2. All-in-one nappies

A shaped nappy, where the padding, liner and cover are all sewn together in a shape suitable for the easy dressing of the baby. Again, they are usually made from cotton – look for organic versions. They usually have Velcro fastenings or similar.

Great for

No folding, no pins, no separate parts to get lost or come undone.

Bad points

Can take quite a while to dry, not usually suitable for tumble-drying, can be quite an exact fit, so you may need several sizes as your baby grows which adds to the cost, need to change the whole thing with every wetting or motion.

How to use

These are the easiest and simplest to use: lay one out flat beneath the baby and bring the sides round and the flap between the legs, fastening them with the Velcro or similar strips attached to them. When using these, it's important to ensure that they fit correctly, especially around the legs where leaks can occur. If the legs are too loose, it means you should either adjust the tapes to bring them closer to the legs or even switch to a different size. If they are too tight, they can cause chafing and soreness around the baby's legs, so move up to the next size.

As these nappies are, as the name suggests, all sewn together, they can take quite a long time to dry. Tempting as it may be to throw them in the tumble-dryer, try to resist the temptation. In addition to the extra energy used and carbon emissions this will create, they can actually have a shorter lifespan if you do tumble-dry them, as the heat and dry air can cause the waterproof outer to crack and perish. Try to line-dry in summer and on less clement days put them over the radiators or on an airer. It is a good idea to use liners, either washable or degradable paper, as if the nappy is dirty but not wet, you can reuse the nappy. Dispose of liners in degradable nappy sacks but try to put a few in a bag before throwing away.

Where to buy
See *Resources*, 4.

3. Two-part nappies

These usually have a cover, with a breathable yet waterproof outer layer, made from cotton and fastenings, usually Velcro. Inside this you place a washable fabric pad, which can be folded in such a way as to be suitable for either a boy or a girl. Some two-part nappies have a shaped, adjustable cotton inner with fastenings and a waterproof pant to put on over the top. Their appeal lies in a few areas – you can adjust the inner pad to be suitable for the sex of your child and his or her nappy needs at the time, you can add extra layers for night time and when baby is particularly wet, and you can reuse the outer if it is not wet or dirty.

Liners are a good idea with this sort of nappy. Two-parters are quicker to dry than an all-in-one and the inner pads can often be used through several sizes, depending on the system you choose, all you need to do is change the waterproof covers to accommodate your growing baby. They are also often slimmer than all-in-ones, making for a neater fit under clothing.

Great for
Easy to get a good fit, covers may be reused for several nappy changes if not wet or dirty, quite quick to wash and dry.

Bad points
You may need several sizes as baby grows, you may have to buy quite a few covers to cope with all the changes.

How to use
Also called a pocket nappy, these are sometimes easier to use than an all-in-one, though may at first appear more fiddly. You simply fold the inner to be suitable for the sex of your baby, tuck it into the outer and fasten the whole thing with the adjustable Velcro or tapes.

Where to buy
Again, look for organic versions first and you will find that there are many different

brands available but do some research to work out which ones will be best for you; ask friends, read the company's literature (some will send you a sample before you decide), look online and ask in your local nursery shop. There are groups who can help with your choice, too, such as your local Real Nappy Association (see *Resources*, 1) representative. See *Resources*, 4 for popular washable brands and where to find them.

One of the best resources is the excellent website and helpline set up by The Women's Environmental Network. They can also offer you details of nappy services in your area.

HOW TO GET THE BEST OUT OF YOUR WASHABLE NAPPIES

Don't buy too many nappies before the birth – the newborn and even first-size shaped nappies are often only used for a short amount of time. Get one or two and have a pal primed to go shopping for you when you give birth and know how big your baby is. Keep all the nappies in their packaging so you can exchange them if necessary.

Washing nappies

Should I use a laundry service?

If you don't want to wash nappies every week, you could try out a nappy service. There are environmental drawbacks – the fact that they use vans to deliver and collect, the detergents they use and the energy required to launder the nappies – but if you really want to try to use washable nappies but find it too hard – for example, if your baby is at nursery and you are working – a laundry service may be a viable alternative.

Nappy laundry services often supply you with all the nappies, wraps and storage bins you need, collecting, laundering and returning them as required. Of course there is a cost involved – WEN research in 2006 found most charged between £6.00 and £11.34 per week[13] – but they are still a relatively green option, due to the fact that the laundry will wash in the most cost-effective, energy- and water-efficient way. In other words, the laundry will be washing full loads of nappies all the time, rather than the possibility of your having a half-load of nappies if you have run out.

Find your local laundry service in The Yellow Pages, through your council or the Women's Environmental Network, or you can get in touch with the National Association of Nappy Services (NANS) (see *Resources*, 5) which not only promotes the use of cloth nappies but protects the public by ensuring that only approved NHS regulations regarding laundry procedures are followed – this ensures the nappies are thermally disinfected, i.e. the heat used to wash them kills germs and bacteria.

ECO TIP: Washing a nappy still uses up to five times less energy than making a disposable[14]

Eco-friendly disposables

Again, there are environmental pros and cons to these. Although, generally speaking, they are made from ingredients that, on balance, are possibly sustainably sourced and better for the environment, they are still a problem to dispose of and are not really suitable for composting, as some of their manufacturers claim. If they biodegrade unaerobically, they produce methane, which as a climate change gas is 21 times more potent than carbon dioxide! So they are a step in the right direction, but not as good as washable nappies.

The benefits

Though exact proportions of the ingredients vary, most of the eco disposable nappies on the market in the UK share these benefits:

- They use ingredients from renewable resources, such as managed forests for the pulp.
- They contain only unbleached cellulose pulp.
- They are 100% chlorine-free.
- With some brands, a proportion of the gel inside them is biodegradable. Others contain no gel at all.
- The plastic outer is biodegradable.
- The packaging may be biodegradable.
- They are latex-free.
- They are breathable, so kind to baby's skin.

What's available?

Some brands use absolutely no gel at all, relying on wood pulp and cotton for absorbency. They are also latex-, coloured dye- and perfume-free, do not use any genetically modified ingredients, and the woodpulp they contain is 100% chlorine-free.

ECO WASHING TIPS

Always wash on a full load.

Wash on as low a temperature as possible – you can make a judgement about how soiled they are! Wet nappies can be washed at a lower temperature (30° or 40°) than soiled ones and 60° is fine for those.

Choose eco-friendly detergents (see Chapter 7 for examples of how and where to buy eco-friendly washing products).

Avoid using the tumble-dryer unless you're desperate! Use the available heat from your radiators, or on sunny days, use the washing line.

Drying nappies outside keeps them whiter and smelling fresh.

THE PROS AND CONS OF DISPOSABLE VERSUS CLOTH NAPPIES

	Cloth	Disposable
Pros	Use less of the Earth's resources Are reusable Are cheaper Are recyclable for other babies Becoming more user-friendly Shaped, Velcro or similar fastenings	Convenient More absorbent Take wetness away from skin
Cons	Need more frequent changing Need pre-soaking, buckets lying around etc. Non-organic cotton ones may have been bleached Need washing	Take 100 years or even longer to degrade in landfill site Lead to later potty-training and so more nappy use Ingredients may irritate skin
Myths	Need washing using strong detergents – you can use eco-friendly laundry liquids or powders, Eco Balls or similar. Need to use high temperature (60°C is adequate)	

Some are made in the US and they have to be shipped over to the UK, increasing your carbon footprint just when you're trying to reduce it. Look for UK or, at the very least, European brands.

Other brands have a small amount of gel for extra absorbency (see *Resources*, 6) combining an eco-friendly solution with practicality – great for the occasions when you need to go a little longer between nappy changes, such as night times and when travelling.

Brands which have no bleaching agents, perfumes or lotions are particularly recommended for babies with eczema or sensitive skin. Another brand boasts biodegradable packaging, and the outer layer of the nappy is also biodegradable.

They can be composted in around eight weeks on your own compost heap, provided you do not put too many on at once. However, they will not turn into compost on landfill sites due to anaerobic conditions.

Others use pulp from renewable forestry sources and they are oxygen- rather than chlorine-bleached. They use no optical brighteners, perfumes, lotions or moisturisers. The core contains very absorbent starch, which is 100% biodegradable (see *Resources,* 6).

Try before you buy

Some web-based companies offer a trial of the different types of eco disposables and some will send you samples so you can try them out on your baby before you buy. Internet stores may also offer free delivery if you order a certain amount, so it pays to buy in bulk.

Nappy accessories

What other pieces of kit do you need? Here are the essentials…

Bucket

A good bucket with a well-fitting lid is ideal for nappy laundry. Use it to store and/or soak nappies. In fact, if you have two, you can use one for dirty nappies, the other for ones that are just wet.

In the bucket, you can put a few different things to help deal with bacteria and odours. Bambino Mio's Nappy Cleanser is an environmentally friendly, biodegradable cleanser with an antibacterial, germicidal action. It also deodorises. You don't need to boil, wash or soak the nappies and it doesn't contain enzymes, synthetic perfumes, chlorine bleach or optical brighteners found in conventional nappy sanitisers. In addition, why not invest in a couple of laundry nets? Use a net to line the bucket and when it's full, just lift the whole thing out and straight into the washing-machine.

Fastenings

Safety pins

The oldest method and still a good one, as they allow you to adjust the nappy exactly to your baby's size. Be sure to get the versions with pull up, snap down heads which prevent the pin becoming undone and always put your hand between the nappy and the baby when fastening them to avoid accidents.

Plastic fasteners

A less environmentally friendly alternative to pins, these are little three-armed gadgets. The ends of the arms have little 'teeth' which hook gently into the fabric of the nappy and hold it in place. They are placed on either side of the waist, holding the back, front and underneath parts of the nappy together (see *Resources,* 7).

Liners

An essential aid to your nappy laundering, you place these inside the nappy where they catch the worst of the mess. They are often the type called 'one-way' liners, which allow moisture to soak through but prevent it from seeping back again, thus keeping baby's skin dry. Make sure you buy degradable or better still, washable ones (see *Resources*, 8).

Nappy bags

Out and about? Wet or smelly washable nappies in your bag? Invest in a couple of washable nappy bags (see *Resources*, 9) or buy a pack of biodegradable plastic nappy bags. Just pop the nappies in the bucket when you get home (see *Resources*, 10).

Nappy Outers, Plastic Pants, Nappy Covers

If you're using terry nappies, you'll need pants to put on over the top to prevent leaks. There are some economical pants available from most nursery stores, such as basic nylon waterproof covers suitable for nights and also breathable ones, great for day-time use (see *Resources*, 11).

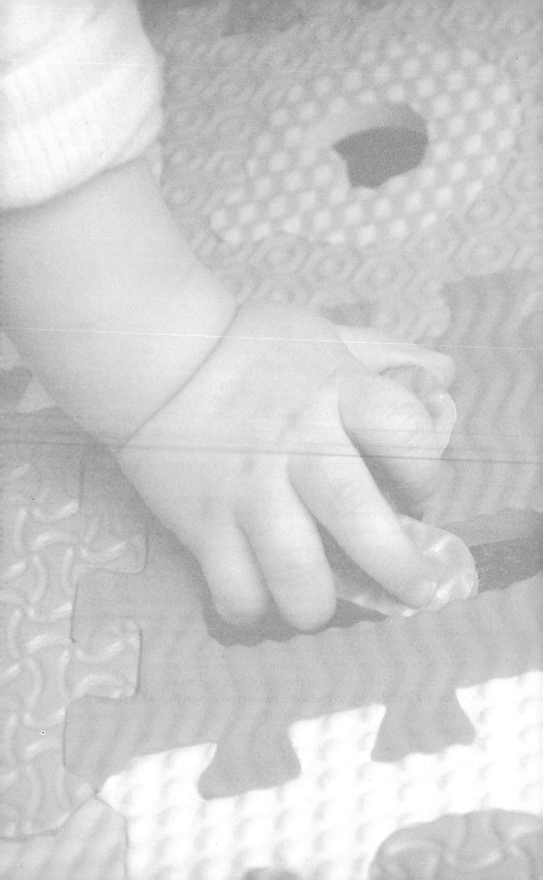

Chapter Four

The Nursery

DECORATING

When a new baby is on the way, you may want to redecorate, especially if you are lucky enough to have a spare room for a nursery. The Foundation for the Study of Infant Deaths recommends that a new baby sleeps in the same room as its parents for the first six months, but making a nice, new room for your baby to move into later is a pleasure.

A lot of the advice in this chapter on the use of solvents and chemicals in paints and other decorating products might make you think you should completely re-decorate your house; however, once you are aware of all the facts you can then choose to do as much as you want to meet your environmental criteria – and of course your lifestyle, budget and time.

When you are pregnant and when you have a tiny baby in the house, with his new, fresh lungs, it is instinctive to want to avoid anything chemical or smelly and things that may give off hazardous fumes in your home. When my daughter was first born, we were in the middle of having our hallway decorated and due to the builders' timetable, it couldn't be put off if we wanted it done sometime that year! I asked the major paint companies what went into their products and whether they were safe to use around a newborn. They all told me that their paints were entirely safe to use and would cause no problems. I wish I had known then what I know now – that although all the ingredients in mainstream paints, varnishes and stains are considered 'safe' by the industry and the Government, they can actually be extremely harmful in more ways than one.

I have a draughty house to thank for the fact that we experienced few problems – plus the fact that it was summer meant we could keep doors and windows open almost all day. However, if I could rewind, I would definitely have looked into the possibility of using eco-friendly products, not just from the health aspect but also because of the effects of the paint industry on the environment.

We are constantly being told by the media that allergies are on the increase. Whilst some argue that this is due to better recognition and reporting of allergies, I am personally in no doubt at all that allergies are indeed becoming increasingly prevalent. When I was at junior school, there was only one child in a school of around 400 children with eczema – in fact, she was known as "the girl with eczema". Now, it seems I know personally of several children with eczema, a number with asthma (which was almost unknown when I was a child) and many more with food intolerances and reactions to certain animals, plants and substances. It seems obvious to me that something is going on and that our modern lifestyle, particularly in cities and towns, is to blame. We are all exposed to a toxic 'soup', caused by carbon emissions from cars and power plants, refuse incineration, chemicals in our household products and in things used for building and decorating our homes.

Fortunately, in addition to paints, varnishes and other decorating products, there are many other things that go into the decorating of a baby's room that, when carefully chosen, can have a positive effect on your baby's health and on the look and feel of the room. Natural fabrics and flooring can make the room look great without introducing harmful chemicals and fumes into it. Mass-produced rugs and carpets made from man-made products can harbour chemical smells and all carpets and rugs can be home to nasties like dust mites.

In addition, lighting, heating and electronic products can create positive or negative effects on the environment and your baby's sleep and health – this chapter will help you to create an eco-friendly room for your baby to play, relax and (hopefully) sleep in.

Eco paint and decorating products

There are now many suppliers of eco-friendly paints, many of them available in independent shops, on the internet and by mail order (see *Resources*, 1).

The ingredients in eco paint

Most 'natural' or 'eco' paints are made using natural ingredients based on traditional methods of paint manufacture. They may contain minerals or plant-based ingredients, including any of the following:

> Beech cellulose, beeswax, borax, boric salts, chalk, citrus peel oil, clove oil, dehydrated castor (stand) oil, earth and mineral pigments, eucalyptus oil, fir-needle oil, glycerin, iron mica, kaolin, lavender oil, linseed oil (refined), marble lime hydrate, mica, milk casein, natural asphalt, natural gum milk, potash, quartz powder/quartz sand, shellac, swelling clay, talcum, thyme oil, wood stand oil.

Apart from the fact that they don't cause toxins to be released during manufacture, the benefits of eco paints are that they are not bad-smelling, they contain no carcinogenic ingredients, they do not provoke allergies and asthma, they are easy to clean from brushes and rollers and they actually help improve air quality as clay-based paint absorbs odours and contaminants.

Several of the same manufacturers who supply natural paints also offer products for wood. These include coloured wax-based stains, wax oils and other products to give subtle colour and sheen to the wood, also providing a waterproof finish (see *Resources*, 2).

Mainstream paints and decorating products

In the past, paint was a very organic product, made from local pigments and materials. During the Georgian and Victorian eras, paint became an important part of interior décor and the range of colours was expanded to suit the grand, opulent homes of the rich. The poor, however, still relied on distemper (powdered chalk mixed with size, a gelatinous glue made from animal bones) or a simple, natural paint to bring lightness and colour into their homes. Painting your house was often a very local affair, as paint needed to be mixed on-site from the available ingredients. To obtain a blue-green, for example, the painters would mix the pigment terra verde, found in many rural areas, with buttermilk and egg white! I wonder what is smelt like as it was drying – or in particularly hot weather.

Today, however, paint is made in huge quantities in chemical factories and ingredients are far from natural. When you read this list of precautions found on a mainstream paint manufacturer's website you might want to think again before driving to the DIY superstore to pick up a can of white gloss:

- Ensure good ventilation during application and drying.

- Avoid contact with skin and eyes.

- In case of contact with eyes, rinse immediately with plenty of water and seek medical advice.

- After contact with skin, wash immediately with plenty of soap and water or a recognised skin cleaner.

- When spraying do not breathe spray, wear suitable respiratory protective equipment.

- Keep out of reach of children.

The major brands of paint, varnish and stain commercially available in this country are considered safe to use in our homes. They are tested and approved by scientists and safety bodies, in the UK and the EU, and have been tested by many different bodies and scientists. Yet the ingredients that are used in them, processes used to make them and the after-effects when they have been applied to our homes all have damaging effects on the environment and on our health – and that of our babies.

Over the past decades, advances in chemistry have meant that scientists and researchers working in the decorating industry have come up with paints and finishes that have improved in the consistency of their colours, the way they can cover a wall and the way they are applied. The palate of colours has increased dramatically as we demand whiter whites and more vibrant colours. To do all this, the manufacturers have added more and more chemicals and non-organic ingredients, many of which have the potential to damage our health.

The ingredients in mainstream paint

Mainstream paint contains VOCs; the term 'volatile organic compounds' [1] was coined by the paint industry to mean the vapour and fumes given off by paint as it dries. The lower the VOC, the safer the paint.

In addition to the pigments, which may be organically or chemically derived, mainstream paint may also contain solvents to keep it runny and help it dry, as well as pesticides and fungicides. These all cause pollution at the point of manufacture and can give off fumes when drying and in some cases for long afterwards. Some gloss paints have the highest levels of VOCs (between 25% and 50% according to the information on one major manufacturer's packaging) and even an emulsion can have up to 8% VOC content. Fungicides and pesticides can cause health problems in humans at high enough doses but you may want to consider whether or not you want them near your baby in any dose.

Wallpaper

In an effort to bring prices down in this competitive market, manufacturers use the cheapest methods when producing wallpaper; this often means that chemicals and artificial inks are used and their waste gives off Volatile Organic Compounds just like paint. If you smell most cheap, commercially available wallpapers, you will notice a strong chemical smell, just as you do with paints.

If you choose papers that have been made from sustainable paper sources and whose colours are created using natural paints, smells and fumes will not be an issue. There are some wallpapers available (see *Resources*, 3) that are made from hemp, which is a fast-growing and natural resource.

Adhesives

Wallpaper paste often contains solvents, acrylics, pesticides and fungicides, all of which are potentially hazardous. Look for a more natural form of adhesive (see *Resources*, 4), which is easy to mix and apply, suitable for all types of paper and contains no 'nasties.'

Fume absorbers

If you have recently painted or varnished using conventional products and are now concerned about the amount of odour coming from your walls and furniture, there are a couple of methods of absorbing fumes and making the air in the nursery cleaner and less toxic.

One is made from sheets of black material impregnated with activated charcoal, which consists of highly porous carbon granules that are free of volatile materials. Its thousands of tiny pores give it a huge surface area and it can absorb odours, gases from cooking, nitrogen dioxide, sulphur dioxide, paint fumes, smoke and ozone. Fume absorbers should be replaced every three months (see *Resources*, 5). Hang out of the reach of babies and children and away from areas where pets could disturb them.

Bicarbonate of soda is good at absorbing smells – leave an open box in the room but well out of the reach of babies and children. There are some odour-absorbing candles on the market, but make sure you put them out of reach of children.

Alternatively, you could consider buying an electronic air purifier, but check the manufacturer's recommendations as to which one will best suit your home and needs (see *Resources*, 6) and unless you are convinced your home is particularly polluted, you should try natural methods of cleansing the air before considering buying a machine which has to be manufactured and transported and which also uses energy.

Removing old paint and varnish

Chipped, peeling paint

The removal of old and damaged paint inevitably involves dust, and a dusty environment is not good for a pregnant woman or a baby; so if you have a lot of work to be done that involves the removal of old paint or finishes, keep the rooms well ventilated or in extreme cases, consider moving out for a few days while it is done.

Lead paint

If you suspect that furniture, windows or other items have been painted with lead paint, extreme care must be taken to remove the old paint or cover it. If the paint is cracked or peeling, it may give off tiny particles of lead which can lead to poisoning. Do not attempt this if you are pregnant or you have a new baby. See the advice on page 22 in Chapter 1 on lead paint.

MDF

MDF leaches formaldehyde which has the potential to damage your health, particularly if the dust is inhaled. Do not attempt to sand old MDF items and ideally consider replacing them.

Paint-strippers

Conventional paint-stripper smells bad and is unpleasant to use. The fact that you are advised to wear gloves and even a mask gives you a clue as to its potential to damage your health – it contains solvents which can, in large enough quantities, cause brain and nervous system damage and not only are the fumes strong and overpowering but the stuff can burn your skin too. There have been reports of the excessive use of paint-strippers causing loss of memory and brain function, heart problems and even short-term diabetes. [2]

Most suppliers of eco paint will be able to sell you a paint-stripper too. Most of them contain no solvents and can remove several coats of paint. They wash off with water and don't affect the wood underneath (see *Resources*, 7).

Hard flooring

Wooden floorboards are an ideal choice for the nursery, and if you are lucky enough to have them, simply sand and varnish the boards using an eco-friendly product. Wood is clean and does not harbour dust mites or give off any fumes.

It is best to have someone professional do the work for you, as old paint or varnish on the floor may contain VOCs and lead which you do not want to breathe in at any time but particularly when pregnant. Make sure, as well, that the work is done well before the birth, to allow dust and fumes to clear. If you need new boards, make sure they are FSC- or PEFC-approved pine or similar.

There are other hard-flooring options but many of them contain glues and other chemicals which may emit formaldehyde – laminate flooring for example, which has chipboard beneath the wooden top. And vinyl, coming from a petrochemical source, is another one to avoid if you're going for green options. Lino – which is a natural product made from sustainable sources – is a good alternative.

There is some bamboo flooring on the market which seems to tread a path between being green and having some elements which may cause concern: whilst bamboo is a plentiful resource and is highly sustainable, the floor may be glued together and finished with chemical-based varnish or stain. Check with the manufacturer to find out what they use. Whilst these glues and stains may give off gases for a while, reports say that the gases last a shorter time than those in laminate flooring. It would seem to be a case of buyer beware. [3]

Soft Furnishings

Carpets and rugs

If you are choosing to re-carpet, look for a good quality wool carpet rather than nylon or another synthetic material. Wool is hard-wearing, natural and is also naturally fire-resistant – it also ages beautifully. Other natural fibres are also popular for flooring such as cashmere, cotton, jute, coir, bamboo and grasses. Make sure they have not been chemically treated and that the backing material is also environmentally friendly.

If you are choosing rugs, try to look for natural fibres and pick rugs that can be machine-washed, using environmentally friendly fabric cleaner (see Chapter 7). Several environmentally friendly suppliers offer rugs in a great range of colours and designs (see *Resources*, 8).

Carpet and rug cleaning

If you are decorating the new baby's room and there's a perfectly good carpet down, only it has a few stains and marks, be aware of the chemicals which are available for carpet cleaning and which are usually used to clean carpets by professional companies.

A quick look at one of the spot cleaners available in the UK lists the following ingredients: surfactants, polycarboxylates, phosphonates, artifical perfumes and preservatives.

Some carpet cleaning firms are starting to offer environmentally friendly cleaning products as part of their service, but do check with them to see what products they use. If they are unable to tell you if the cleaning fluids are organic or chemical in origin, find a firm that does know! Alternatively, use a natural environmentally friendly cleaner in your aqua vac or a spot cleaner which can remove stains from clothes, carpets, fabrics, upholstery, leather, plastics and more (see *Resources*, 9) or use one of the natural, old-fashioned cleaning suggestions in Chapter 7.

Fabric, curtains and blinds

Lightweight curtains made from natural, organic cotton or other natural fabrics make a good choice in the nursery (see *Resources*, 10). If you find that they do not keep out enough light and your baby is waking up really early, line them with thicker fluffy material such as brushed organic cotton, or if you are having curtains made for you, ask for this kind of lining. They should be washed frequently to prevent the build up of allergens.

Alternatively, a thick blind can keep the room nice and dark. Wooden blinds are a good choice ecologically, though it is important to keep the slats clean and dust-free, which can be a fiddly chore.

Sofas and cushions

Introducing a seat or sofa and cushions into the nursery makes the room cosy and restful, plus it is more comfortable for you during those long night feeds. However, they can also provide areas in which dust mites and dust can accumulate, so you may want to avoid too much soft furnishing. Plain wooden furniture makes a better choice, particularly if there is a family history of asthma or allergies. Choose cushions made from natural fibres and wash their covers frequently to prevent a build-up of allergens.

ALLERGY ADVICE

It's much easier to keep the baby's room clean and free of mess, dust, dust mites and other allergens if you cut down on the amount of fabrics, carpets and rugs in the room. Wooden floors with a rug or two and smooth, easy to clean surfaces will prevent the build up of dust and mess and make a pleasant, restful environment for your child to sleep in.

Lighting

You can use lighting to create a calm, peaceful environment in your baby's bedroom. Try to keep it light enough to be able to see all you need to when getting up for night feeds and changes, but keep the lighting soft rather than harsh.

In the day-time, of course, you can make use of natural light, so make sure curtain poles are long enough that when the curtains are open they do not block light from the sides of the window.

Use energy-efficient bulbs in the room, as well as in the rest of the house – they last as much as 12 times as long as conventional light bulbs, are available for all kinds of lamps, and many can now be used with dimmer switches, (check before you buy if you want to be able to dim the light).

A recent innovation in low-energy bulbs is the BioBulb (see *Resources*, 11) which uses 75% less energy than a standard bulb, is bright and mimics sunlight, which may improve mood, night-time sleep and daytime energy. It lasts around seven years and has a pleasing white, flicker-free light. It makes a great choice for a nursery and indeed for your own bedroom. BioBulbs are quite bright, so a couple of dimmer side lights – with low-energy bulbs of course – will help you and baby relax in the evenings.

Many parents think a night light is essential in the baby's room and keep it turned on all night, which wastes energy unnecessarily: it helps with the sleep patterns of a new baby if she can learn from early on to distinguish between day and night. So the rule is, light and noise is fine during the day but night-time must be dark and quiet. Don't make the bedroom so dark for day-time naps that they think it is night – and by the same token, don't keep the night light on all night or they will not know what time it is! This is confirmed by scientists who have studied sleep patterns. Our brain has a 'clock' that determines our sleeping behaviour and whilst light makes the brain think it is day-time, as it gets dark, the brain releases melatonin which helps us know when we need to sleep.

If you must use a night light – for example, in the hallway to help you get to baby's room without falling over – look at the energy-saving options. You could choose the new Moonlight Nightlight, which uses NASA space technology to run the light for 24 hours a day, all year for under 50p (see *Resources*, 12). It has no bulb but an electro-luminescent panel that is cool to touch and emits a blue/green light.

Electronic devices

Try to avoid electronic gadgets, especially stimulating ones like television, in the nursery. They are usually unnecessary and can cause problems when you try to get the baby to sleep. If babies are exposed to stimuli just before bedtime, it can take them longer to settle and it can interfere with sleep patterns. To keep the room peaceful and calm, try not to have computers, telephones, mobile phones or televisions in the room. If this is unavoidable, for example if you have a home office in the same room as your baby, then turn everything off unless you are actually using it. There is some evidence that the humble Spider Plant and the Peace Lily can purify the air and can remove toxins.

You may want to consider the possible health issues around electronic devices in the nursery. Some scientists [4] believe that microwaves, electromagnetic fields and the emissions from mobile and cordless phones, wireless computers and other devices, can interfere with our brain patterns and lead to health problems such as tiredness, lack of concentration, depression, psychological problems, insomnia, lethargy and sometimes tingling in the fingers. If you are concerned about electromagnetic fields, remove electronic, especially remote control or cordless, devices from your baby's room or at the very least turn them all off when you are not using them.

For more information on this subject, see *Resources:* Publications.

Chapter Five

Clothes

BABY CLOTHES IN NATURAL FABRICS

It is becoming increasingly easy to buy environmentally friendly, organic and ethically sourced baby clothes; shops, mail order catalogues and Internet sites selling clothes that are made from organically grown fibres, from natural fabrics and from sources that promise the workers are treated fairly have increased exponentially over the last few years. Be aware that there are several different terms in use to describe ethical clothing – see the introduction on pages 10 and 11 for the differences between organic, ethical and fair trade.

Your choice of clothing can make a difference to your baby's general health and comfort; as clothes are worn next to the baby's skin, which is five times thinner than that of an adult, you need to be sure that they are as natural as possible and free of chemicals and ingredients which might cause irritation, rashes and allergies, like eczema and psoriasis. Both conditions can improve dramatically as long as the skin is kept cool, is able to breathe and is not irritated by chemicals and artificial dyes. Artificial fibres, particularly nylons and polyesters, may be great at keeping us warm in winter and may wash and wear well, but they trap heat against the skin and do not allow the baby to regulate his own temperature efficiently.

Organic cotton

Many ethical companies and some of the high street names (see *Resources,* 1) are now providing baby clothes made from soft, cosy organic cotton – and as more of us choose it, the price is becoming more affordable for everyone. So what makes organic cotton an environmentally better and more ethical choice?

- Organic cotton is grown with no chemicals, pesticides or herbicides.

- Its farmers use no chemical fertilisers, making use instead of natural manures and crop rotation to replenish the fertility of the soil.

- The final cloth is normally not chlorine-bleached.

- It is generally coloured using natural, plant-based dyes.

- All production of ethical organic cotton cloth – weaving, spinning, washing etc. – is carried out responsibly.

- Organic farming works with nature, leaving hedges and field edges that provide habitats for natural predators like beetles and spiders, which control pests.

- Organic farms have 44% more birds in the fields and five times as many wild plants as traditional industrial farms.

- Organic cotton is not only soft and thick, it is more durable than chemically farmed cotton, and garments made from it will last longer.

Bamboo

- Bamboo is a great source of environmentally friendly fibres:

- It grows incredibly fast, reaching its full size in just four years.

- It uses little in the way of nutrition and yet enriches the soil it grows in.

- It can be grown with no chemicals or pesticides.

- It is 100% biodegradable.

- it regenerates itself naturally, sending out new shoots from its roots, which multiply the size of the plant naturally.

When woven, bamboo has a subtle sheen and feels as soft and light as silk or cashmere. It also has natural anti-bacterial properties, which help keep smells at bay – fantastic when used for nappies! It is a good thermal regulator, keeping you cool in summer and warm in winter. Most of the supplies coming onto the market have been ethically manufactured, plus many are also fairly traded. Some manufacturers are using bamboo for sleepwear and also for beachwear, for children and adults alike. It is UV-resistant and wicks moisture away from skin, keeping you and baby comfortable, protected and happy (see *Resources*, 2).

Hemp

Although hemp is more often associated with drug production, a variety of the hemp plant can be used for many different purposes and it is an excellent environmentally friendly source of raw material for organic cloth:

- It is a totally renewable resource.

- It grows easily and fast and suppresses weeds.

- It produces 250% more fibre per acre than cotton and uses far less water.

- It has long fibres that get softer each time the garment is washed.

There are now a few suppliers of hemp baby clothes in the UK. You may be able to find suppliers of fabric by the metre, which you can make into clothes, bed linens and curtains, if you're handy with a needle (see *Resources*, 3).

Linen

Made from the fine, long fibres of the flax plant, organic linen is another good choice for baby clothes, though it is a bit more expensive than some other fabrics, a point to consider when you realise how short a time your baby will wear an item for. Linen has traditional been grown and woven in northern Europe and pre-dates cotton in these areas as the original nappy.

ECO BABY'S WARDROBE – THE BASICS

This section will give you advice on the clothing you need for a newborn. You will doubtless add other items and be given clothing but this gives you the low-down on what kind of clothes I found most useful when my kids were newborns.

> ECO TIPS: Keep all clothing for your newborn in its packaging and keep the receipts; if your baby is smaller or larger than expected or has an unusual growth spurt, you can exchange them or get a refund.
>
> Don't buy too much of the first-size clothes, as most babies outgrow them really quickly and you may be given clothes as presents.
>
> Start small and buy as you need things.

Vests

I can't praise vests enough – that is, the kind that does up under the crotch with poppers. They keep nappies in place and allow for easy changes, plus they keep the baby cosy. They come in short- or long-sleeved versions or even sleeveless, so you can choose the appropriate ones for the time of year. In the really hot days of summer, this is all a baby needs. Choose several organic cotton vests like this for each stage of growth and you can't go far wrong. For the very early months, the ones with an 'envelope' neck are really good, as you may be nervous of dressing such a tiny one and it helps get the vest over the head with no awkward, stuck moments!

Baby grows

Also called babygros and onesies, these are the all-in-one body suits with feet attached and long sleeves, fastened up the front with poppers. You may come across versions where there are poppers or even buttons on the back, but as all advice is to lay small babies on their backs, this is not only uncomfortable but also really fiddly when changing them.

Choose organic cotton or organic cotton towelling – it is soft and comfortable next to the skin. Apart from special occasions and going out, your baby can practically live in these. If your baby's nails are sharp or she/he has eczema, look for ones with a little fold-out glove on the end of the sleeve.

Cardigans

A soft, pure organic baby wool or organic cotton cardigan makes a great cover-up for cooler days. The best ones are front-fastening – they may have a wrap-over front like a kimono, or buttons or poppers down the middle.

Outerwear

Soft hats and gloves in organic cotton or wool are great – though be prepared for the fact that babies throw them off on a regular basis! Avoid polyester fleece as although it is cheap (and you may get fed up with buying new hats and gloves), it does not allow the skin to breathe and the baby can get really grumpy. Polyester is a petrochemically derived man-made fabric too, so worth avoiding on both counts.

A padded jumpsuit is great for the winter as opposed to a coat, as it keeps the baby warm from top to toe. The smaller sizes often have the feet attached but as they get bigger they tend to be footless to allow you to put shoes on the baby. Choose natural fabrics which will both be environmentally friendly and ensure the baby can regulate his temperature well and not get too hot.

Sleepwear

You don't need anything too fancy for a baby's nightwear and although you'll find some companies who offer cute nightdresses and miniature pyjamas, a baby grow works just as well for night as for day. Having said that, I was completely converted to the sleeping-bag when I had my second child. A concept that has been around in Europe for ages, sleeping-bags were just hitting the UK when my little girl was born. The concept was so simple that I tried it, and she slept beautifully. There is now a great choice of sleeping-bags in organic fabric (see *Resources,* 4).

As the name suggests, it's just a sleeping-bag shape with a kind of vest top, made from padded cotton or cotton jersey. Some fasten on the shoulder, some down the front or some around the side. For changes at night-time, the ones with bottom access are best as you don't have to get the baby out of the whole thing just to change a nappy.

Underneath, the baby wears a long-sleeved top or vest and you can adjust the layers according to the weather – add a baby grow for cold days maybe a cotton cardigan for very cold ones and just the vest when it's warm enough. To allow for night-time accidents and explosive nappies, it may be best to buy two, and even though they are quite pricey the sleep benefits more than make up for the cost.

ALLERGY AND OVERHEATING

For babies that have eczema and other skin conditions, try to make sure that every garment touching the skin is

- As natural as possible.

- A good thermal regulator.

- Not too hot, as this can make the problem worse.

- Easily removed so you can regulate the temperature for changeable days.

- Has not been chemically treated, for example with chlorine bleach and formaldehyde.

- Has not been washed in detergents, which might irritate further.

ETHICAL CLOTHING

The message is starting to get through

The demand for ethically certified cotton doubled over a six-month period during 2006, according to the Fairtrade Foundation. In addition, the sales of fair-trade cotton rose by 4,000 per cent in volume and over 300 per cent in value over 2006. From just seven producers who qualified for the Fairtrade logo in 2005, the number has now risen to 30.[1]

Over the last ten years, particularly if you go back to the beginning of that period, the clothing suppliers who offered organic material and fairly traded products tended to be small companies, often regarded as a bit 'hippy' – and their clothes tended to reflect that look too, with a choice of either garish colours or drab greys and browns and not much in between.

Since then, things have changed a lot. Today, we are not only able to find clothes that are modern and fashionable made from these ethical fabrics, but we are finding that designers are leading the way in making fair-trade and organic clothes really desirable fashion items on the high street. Lots of beautiful, organic and fairly traded baby clothes are appearing and are available in more and more outlets (see *Resources*, 5), making it easy to be a green parent.

There are also several companies who make cruelty-free, animal-free and sweated-labour-free footwear (see *Resources*, 6).

Man-made fabrics

Brief history

In our world of easy-care fabrics and non-iron shirts, man-made fibres have played a huge part in clothing production over the last 50 years and have made the lives of the modern homemaker much easier. The advances in creating fabrics from petrochemical sources exploded after the Second World War and ensured that cloth prices came down and that supplies were plentiful. Synthetic fabrics are easy to care for and have enabled manufacturers to use dyes and weave patterns to create a huge range of patterns in vibrant colours. Clothes can now be made with 'intelligent' fabrics which can, for example, wick moisture away from the skin to keep us cool and dry when exercising, give us lightweight yet waterproof fabrics, thin yet warm skiing clothes – the list is endless.

Nylon – an industrial masterpiece

In 1938 a team of scientists at American chemical company Du Pont mastered the technique of making a thin, strong, yet flexible thread from petrochemicals. Initially used for toothbrush bristles, by 1940 it had become one of the most in demand materials on earth in the form of the nylon stocking, made even more desirable due to their scarcity during the war. Though acetate, a fibre made from cellulose, had appeared earlier in the 1920s, the success of nylon was so dramatic that it was soon followed by several other petrochemical-derived synthetic fabrics such as polyester, rayon and acrylic.

Unforeseen disadvantages

Despite the lessons learned from the appalling pollution caused by coal-powered textile mills and other forms of industry during the Industrial Revolution, the implications of the industrial pollution that would be caused by this new petrochemical industry seem to have been at best, unforeseen and at worst, ignored – possibly because the benefits to the general population – offering cheap fabric to all – were so great (and there was a great deal of money to be made). Huge amounts of capital and manpower were thrown into creating the synthetic fabric industry, which now stretches worldwide.

One of the main problems associated with synthetic textiles is water pollution, caused by discharge from the manufacturing, treating, bleaching and dyeing processes. Oil from machinery may also be emitted. This dirty, contaminated waste is sometimes poured out into rivers and streams untreated, especially in the developing world, often at high temperatures. The combination is dangerous and destructive to plants and animals living in and around the factories and it means that the water source is polluted for those living nearby and relying on it. Noise pollution is another problem, caused by the hundreds of looms involved, as is air pollution for the workers from dust and lint.

Of course, many of these factors are just as true for natural fabrics, when manufactured in large, industrial processing plants, as they are for the synthetic versions. However, with synthetic fabrics the environmental problem is increased by the use of huge amounts of heat, chemicals and water used just to make the fibres in the first place.

Cotton

Non-organic cotton is one of the most environmentally damaging crops in the world as it is sprayed routinely and repeatedly with pesticides.

Mainstream (i.e. non-organic) cotton production

- Because they are not used for food, cotton plants can be sprayed more than food crops.

- In developing countries, 50% of all pesticides purchased go onto cotton fields.

- Only about 10% of the pesticide does its job. The rest runs off into the air, soil, plants and water.

- The pesticides cause widespread water pollution and damage to wildlife.

- In developed countries like the USA, cancer rates are significantly higher near cotton-producing areas. The US Environment Protection Agency considers seven of the pesticides used on cotton to be 'probable' or 'known' carcinogens.

- Farm workers using the pesticides can become ill too.

- After weaving, the fabric is frequently chlorine-bleached causing more industrial waste and damage to water sources and can also be treated with formaldehyde, a known carcinogen.

- Cotton is also a very greedy plant, using 142 pounds per acre of fertiliser, usually artificial.

- The cotton grown for just one t-shirt can use up to 150g of chemicals including Paraquat and parathion.

- Though cotton-growing takes up only 2.4% of the world's cultivated land, it uses 25% of all pesticides and 10% of herbicide produced worldwide each year.

- Chemically treated cotton upsets the balance of the skin by trapping heat and stopping it from breathing, which can cause rashes and eczema.

Remember that babies should not be over-dressed, as this may cause over-heating (see Chapter 1, page 19). The rule of thumb should be that a baby will be comfortable wearing as many layers as you are, plus one extra. So, for example, on a mild summer day, if you are comfy wearing a t-shirt and cotton trousers, the baby will be happy with a similar outfit plus a vest or a light cardigan. A light blanket may be added for naps.

Chapter Six

Food

This chapter aims to give you all the information you need to ensure both you and your baby have a healthy, environmentally friendly and ethical diet, from pregnancy through early feeding to weaning.

So often during pregnancy we are told what foods to avoid, yet many expectant parents want to find out about the best foods for them to eat to increase their baby's health – they don't simply want to be told to eat "a healthy diet". Your diet whilst pregnant is of particular importance to your growing foetus, as is the diet you choose for yourself when breast-feeding and, of course, the products you offer your growing baby as you start to wean. If you are also interested in eating in an eco-friendly and ethical way, you will want to find out about suppliers of organic, healthy and fairly traded foods to help you improve both your diet, the health of the planet and your conscience!

FOOD MILES

Buying food which has travelled miles, often in an aeroplane, is neither eco-friendly nor best for you and your baby; the longer food is stored, the more nutrients are lost, so choose food which is local and in season. That may mean no strawberries in January, but you'll find English strawberries bought in July taste fantastic, unlike the large, watery lumps grown in greenhouses out of season in another country. Avoid products that have been flown in from countries thousands of miles away, like Chilean blueberries and Kenyan asparagus – seasonal food from a local source not only avoids excessive food miles, it tastes better!

FAIR-TRADE

Fair-trade products are really starting to take off. At one time, it was just bananas, then coffee and now it seems that fair-trade products are starting to appear everywhere – and that's great news for the growers and producers of these products, in the Third World and places where labour laws are lax and exploitation rife. When choosing food for yourself whilst pregnant, or later on when you start to introduce first foods to your baby, try to choose fairly traded products where necessary – this is explored in more detail on pages 10 and 92-3.

ECO TIP: Environmentally friendly shopping

- Buy organic – and if you have trouble finding it, pester your local shops to stock more

- Shop locally to avoid food miles

- Buy locally grown produce – it will be fresher, with more nutrients than produce that has travelled a long way, and it avoids food miles – if you can't get British grown fruit and veg, look for the closest European countries.

- If local products are not available, e.g. bananas, buy fair-trade

- Buy what's in season in the UK, avoid produce that has been flown in from overseas and grown using large amounts of fertiliser

- Read labels – know where your food comes from and how it was produced

---

ORGANIC

Why choose organic food?

When I was about eight years old I spent a weekend with my grandmother, who had a wonderful greenhouse full of geraniums and tomato plants. For lunch one day, we had a very simple meal of cheese, bread and one of those tomatoes – mis-shapen, knobbly, slightly split on one side – but with the most wonderful taste! For many years I thought this was merely nostalgia. However, eating my first organic tomato, about ten years ago, brought that day and the greenhouse rushing back – it was simply that for years I had been eating mass-produced fruit which had been grown for conforming size and shape rather than taste. Happily the availability and consumption of organic food, once thought to be the preserve of the trendy, well-off middle classes, is now becoming far more widespread.

Is organic food better for you than conventionally grown products? And by 'conventional', I mean food that is grown using pesticides and artificial fertilisers, with crops chosen for their ability to grow fast and provide high yields, and cultivated using large-scale fields and mechanisation.

Over the last 50 years farming has changed enormously from the small, single farmer-owned businesses that were prevalent before the Second World War (certainly before the Industrial Revolution), to the agribusinesses of today. The way our food is grown or reared has completely altered, as farmers try to ensure they provide large amounts of food at low prices – a benefit in some ways, allowing food prices to drop so that people can afford plenty – but it has been, many might argue, at the expense of quality and taste.

So where do we go for good, scientifically proven information on the pros and cons of organic versus conventional farming? A glance at the website and literature of the Food Standards Agency (FSA) – the Government's department dealing with all issues relating to the food we eat – may leave you thinking that organic farming is a nice pastime for those farmers who want to pursue it and that eating organic food might make you feel virtuous, but has no benefits one way or the other. [1] The FSA says that it is "neither for nor against organic food." On the subject of whether organic or conventional foods are safer to eat, it responds with "Both organic and conventional food have to meet the same legal food safety requirements." It states that it ensures that levels of pesticides "do not pose an unacceptable risk to human health or the environment, and that any pesticide residues left in food will not be harmful to consumers."

But what is "unacceptable"? And why would the Government be so noncommittal about a topic that is so contentious? Farming, though only accounting for around 2% of our country's annual income, is a highly efficient and mechanised industry, providing around 60% of the food needed for home consumption. That makes it quite important in business terms in the UK, contributing £5,238 million to the UK economy in 2005 alone. [2] If the Government or the FSA were to applaud organic farming and suggest that conventional farming methods were producing nutritionally inferior food, there would be a lot of very unhappy farmers and a lot of very confused consumers.

There have also been many misleading stories in the media recently that have said that organic farming gives no greater benefits to the countryside than industrial farming. What stories like this fail to take into account are a whole range of issues apart from land use and the cost of food production. To balance the information provided in the official statements, it is necessary to look at other reports and consider other factors.

There is a lot of factual and scientific evidence [3] that organic fruit and vegetables are better for you in several ways – they have higher vitamin levels, more minerals and of course fewer or no pesticide traces. In addition, the plants chosen are grown for their taste and natural resistance to pests rather than for the ability to produce huge quantities of grain, vegetables or fruit with very little taste.

Vitamin and mineral content

Research has shown that organic fruit and vegetables are higher in levels of antioxidant vitamins, the vitamins most credited with being able to fight cancer, so an organic diet may help you ward off that as well as other illnesses. The plants produce these vitamins to fight off insects and other plants that might compete in their environment.

The Food Commission, a non-governmental campaign group, conducted a study of the nutritional values of the food of the UK between the 1930s and 2002, which was reported in *Food Magazine* in 2006. They discovered that the quantity of essential minerals (in particular calcium, iron and copper) in milk, cheese and meat fell during that period by as much as 70%. The report lays the blame for this at the door of intensive agriculture, in which plants and animals are raised for their quick growth and high yield only. It also blames the poor quality of the soil, resulting from the use of artificial fertilisers, rather than from the traditional practice of returning nutrients to the soil using crop rotation and organic manure.[4]

Toxins in pesticides

A study in Atlanta, USA, which analysed the urine of children aged between three and eleven who ate only organic foods, found that they had virtually no metabolites of malathion or chlorpyrifos, two frequently used pesticides.[5] They then asked the children to return to eating 'conventionally' grown food – and within weeks these metabolites climbed dramatically.

Another study in 2003 at The University of Washington discovered that organically fed children had concentrates of pesticides six times lower than those on conventional diets.[6] Even if these traces of pesticides have no health implications, do you feel happy knowing that what are essentially poisons used to kill bugs and mites are being eaten by your baby with every spoonful of food? Organically produced crops, fruit and vegetables are grown without or with very little use of pesticides or artificial fertilisers, which means that your body – and your unborn foetus and your baby – is not exposed to greater levels of potentially harmful chemicals.

AT A GLANCE: THE BENEFITS OF ORGANIC FARMING

- Organic farming restricts the use of harmful pesticides that may leach off into surrounding plants and water, causing potential health problems for the animals that live there, and polluting watercourses.

- Organic farmers do not use artificial fertilisers, but make use of natural manures and crop rotation to put nutrients back in the soil.

- Animals are reared without the routine use of antibiotics, drugs and wormers used in mainstream farming, thus ensuring their meat and milk does not become contaminated with these substances.

- Field edges are left wild and hedgerows are allowed to grow, providing habitats for wildlife and for insects, bugs and birds that naturally prey on pests.

- Organic farms do not use genetically modified plants, the long-term effects of which we simply do not know.

- Organic food tastes better.

If you would like to find out more about the standards for organic farming in this country and the science behind it, the UK's main licensing body, The Soil Association, is a great source of information (see *Resources*, 1).

BREAST MILK – PERFECTLY FORMULATED FOR YOUR BABY

The importance of your diet during pregnancy and whilst breast-feeding cannot be stressed enough. It gives your baby the very best start in life and in addition there are a number of benefits to breast-feeding quite apart from the nutritional ones – the immediate availability of the milk (no night time bottles to make up), the bonding between mother and baby and the fact that breast-feeding can guard against the chance of your developing breast cancer later in life.

Breast milk has all a growing baby needs for the first six months of life. It contains fat, protein, lactose, vitamins, minerals and water and the proportions of all these change as the baby grows and develops. It is for this reason that unaltered cow's milk is not a good substitute, as the proportions of ingredients vary in all mammals – for example, a seal's milk is very high in fat to keep the baby seal warm, whilst a cow's milk is high in protein so the calf can double its size in a few weeks. You wouldn't want your baby to do the same, though!

To ensure your breast milk is of tip-top quality, you will need to ensure you have as good a balanced diet as you did during pregnancy, with a variety of fruit and vegetables to ensure that your vitamin intake is sufficient for both you and your growing child. If you continue this for the first few months when you are just giving the baby milk and then also when you start to wean, your baby will be as healthy and as eco-friendly as you are!

What to choose if breast-feeding fails

With the best will in the world, not all of us are able to breast-feed – or if we are, we may not be able to continue to feed as long as we would like to. If you are having problems, there are some excellent places you can turn to – for example The National Childbirth Trust has a breast-feeding helpline where you can chat with an experienced adviser who can give you some tips over the phone or give you contact details for breast-feeding clubs or advisors in your area. [7]

Stopping full-time breast-feeding is often linked with returning to work and the problems of turning the baby's feeding over to a childminder or nursery. So if you find that you are not able to breast-feed, there are other options available that do not include buying mass-market, non-organic baby milk powders.

FIRST AND FOLLOW-ON MILKS

There are a number of baby milks on the market, most of which are made by big companies but there are some great organic products, and though they may be slightly harder to find, it is well worth the trouble.

Organic baby milks have been developed to be as digestible as possible and contain two sources of carbohydrates. They have a good protein content and essential fatty acids derived from different cold-pressed organic vegetable oils to provide a source of Omega 3 and 6.

Several companies offer first-stage and follow-on milks with a totally organic, or 99% organic content; the big companies are starting to catch on too, with some of them now offering organic milk alongside their mainstream products (see *Resources, 2*).

LCPs – essential?

Long chain polyunsaturated acids or LCPs are found in breast milk and are added to many baby milk formulas. You will sometimes find them described as Omega 3 oils and the source for these as an additive is usually oily fish.

LCPs occur naturally in breast milk and promote brain, eye and nerve growth. The brain is growing fastest within the first few months of life. Most baby milk companies add LCPs to formula milk, with the exception of organic milk formulas. There are currently no sources of organic LCPs in this country, so organic baby milk will not contain them. LCPs are needed at the time when babies are not being weaned, i.e. in the first six months (WHO guidelines). They are promoted as being essential for a baby's development; however, whilst there is evidence that it is important to give them to premature infants, babies who go to full term and who are breast or bottle-fed are usually able to assimilate them perfectly well from other dietary sources – such as vegetable oils – once they are being weaned.

If you ensure your baby has a good, mixed diet once you start weaning, this should not be a problem. However, if you are in any doubt about this, speak to your GP, Health Visitor or a dietician. If you are breast-feeding, you can obtain these from oily fish or, if you do not eat fish, from other vegetable oils such as flax-seed oil.

Alternatives to dairy milk

If you find that your child has an allergy to dairy products or has a sensitive stomach and finds them hard to digest, or you are vegetarian, vegan or have other dietary restrictions, there are some great organic alternatives available to you:

Soya milk

A useful cow's milk substitute, although Government advice is not to give this to babies under six months – and then only on the advice of a health professional, to avoid the risk of sensitisation to soya protein and exposure to phytoestrogens (information from BDA paediatric group).[8] You can offer first and follow-on soya milks (see *Resources*, 3) after six months, though vegetarians should check the ingredients as one or two contain pork-based proteins. You should also take care to choose only organic soya milk products, as there are more and more genetically modified soya products creeping onto the market. In addition, some farmers are clearing large swathes of virgin rainforest to grow soya for its high yields in a short space of time – an environmental nightmare. See pages 90-91 for more information on genetically modified organisms.

BABY FOOD

The World Health Organisation recommends that you start weaning as close to six months as possible. When you start to introduce cereals and then fruit and vegetables into your baby's diet, it continues to be important to ensure you get the best you possibly can and offer as much organic food as possible to your growing eco baby to ensure he grows healthily!

Making your own baby food

Where possible, make your baby's food from organic, local produce which is in season and is as fresh as possible.

Some of us take to making baby food like a duck to water, puréeing like mad and freezing little ice-cube trays full of tempting treats like butternut squash and spinach. Once you've got into a routine, it can be fairly simple to keep your baby's taste-buds tingling, especially if you use the foods you are giving the rest of the family (allowing for restrictions on some foods according to the baby's age). Remember that you want to make your workload easier, not harder, so try to get your baby interested in what you and the others you feed enjoy – and sooner or later, they will start eating like a 'normal' human being!

Buying ready-made baby food

Whether you buy the occasional jar when you are out and about or whether you intend to feed your baby totally from ready-made meals, there are increasing amounts of really good organic foods available to you (see *Resources* 2). From freshly made and frozen meals to jars packed with organic goodies, meal times can be fun and varied. Ready-made baby food can help you to add variety to your baby's diet and to introduce flavours and textures without having to go to the expense of buying a large quantity of produce. For example, you may not want to buy a whole mango if no one else in the family likes it, but a jar of mango purée might be a winner with your little one. So look for the best and see your baby enjoy her food!

The Food Standards Agency gives Government advice on healthy eating during pregnancy (see *Resources* 4), or you can speak to your midwife or doctor, especially if you have health problems that may cause complications. The NHS also provides advice on what foods may be dangerous in pregnancy and what to avoid via leaflets and its website (see *Resources,* 5).

ORGANIC FRUIT AND VEGETABLES

Many health food shops stock organic fruit and vegetables as well as dried and tinned produce and greengrocers are increasingly stocking local and organic fruit and vegetables.

Farmers' markets are another good source of organic food but be wary of stalls which offer 'naturally' grown food or use other euphemisms that imply organic but are not. Farm shops can also be a good source. In addition, there are also many online organic food companies who deliver to your door – see box schemes, below. You will be given what's in season and can also choose from other produce, including meat, fish and dairy products, that the company may offer. Supermarkets are offering more organic produce than ever before (much of it from abroad – so buy locally grown or British if possible) and most offer a home delivery service. If your local one does not have many organic options, ask them to stock more!

Box schemes

There has been an explosion of organic box schemes in the UK. Most areas of the country are now covered by companies who will deliver a weekly box of seasonal, organic and locally grown produce to your door (see *Resources*, 6).

ORGANIC MEAT

Beef

To be certified organic, beef cattle have to be fed on a natural diet (grass, hay and silage) that is at least 95% organic, and they must be kept outside and allowed to graze freely in all but the very coldest weather. Their pastures must not have been treated with artificial fertilisers or chemical sprays.

There have been no cases of BSE in animals reared on organic farms. The natural diet and lifestyle of organic animals pays dividends in taste, too.

Other meat

Lamb, mutton, pork, venison, wild boar, game birds, goat and even some more exotic species are available either free-range, organic, or additive-free (meaning that the animals' feed has no artificial ingredients but may not be 100% organic) – see *Resources*, 7. All the producers have to adhere to the same standards laid down by the Soil Association.

ORGANIC AND FREE-RANGE CHICKEN

You have choices if you want chicken that is reared humanely and has fewer additives and chemicals. Try to ensure that you know as much about where your chicken and eggs are coming from as possible, either by reading the labels or by asking your butcher or grocer. All eggs should be labelled with their origins – if they are not, the retailer is breaking EU law. If you'd like more information on these issues, visit Compassion in World Farming's website. [9]

Organic chicken

Organically reared chicken has to adhere to the same guidelines as that for beef, i.e. their feed must be organic and they must have good living conditions. Organic birds are kept in buildings where each bird has plenty of space and they are allowed to roam outside during the day, too (see *Resources*, 8).

Free-range

These are chickens that are allowed unlimited access to outdoor space in order to scratch for food, interact with other birds, preen, and exercise. They may have shelter in which to sleep over night and where they may lay their eggs. Some are kept in semi-woodland conditions.

> ECO TIP: To be sure the chickens you get your meat and eggs from are as happy and healthy as possible, go for free-range, or free-range and organic if you want the very best. It tastes better, the birds have had a good life and you will be healthier too.

Additive-free chickens

These have the same living conditions as free-range birds and their food must not contain artificial ingredients but it does not have to be 100% organic.

The chicken industry – what to avoid

A conventional chicken farm – the word 'farm' makes it sound misleadingly pastoral – has huge sheds with literally thousands of chickens, two per square foot, living under constant light. They have no room to move around and routinely have their beaks cut off to prevent the fighting caused by frustration and stress. They are fed growth-promoters and grow so fast that they are often unable to stand and are forced to lie on the litter beneath them, soaked in their own urine, which causes burns on the skin of the legs and breast. They reach their full size in seven weeks – a process which takes twice as long if reared naturally. Because of the poor conditions, they are routinely given antibiotics to prevent the spread of infection.

Despite the antibiotics, many health scares relating to consumption of meat (such as salmonella) can be traced back to battery chickens. Fortunately, new laws are coming in to prevent the use of so many antibiotics, as fears that resistant bugs will proliferate have been fuelled by new health epidemics like avian flu.

There are two categories of bird that live in these conditions:

Barn hens

Don't confuse the term 'barn' hens with free-range– 'barn hens' conjures up images of good conditions, but this is not always the case – the birds may not actually be in cages, but can be packed into an overcrowded barn with very little room to move.

To be certified free-range, the hens have to live under stipulated conditions, as above.

Broilers and battery hens

Broilers (chickens raised for meat) are usually housed in huge barns or sheds, packed in and unable to move about, and egg-laying hens (battery hens) are kept in little cages with barely enough room to move – each hen has a space about the size of a piece of A4 paper.

What's the problem with 'conventionally'-reared animals?

There are several health issues associated with eating conventionally farmed meat. Most of the concern comes from the practice of routinely dosing the animals with antibiotics and other drugs such as growth-promoters, and hormones that are used to increase speedy growth, promote milk production in cows and prevent illness. These can be absorbed into our bodies when the meat is eaten and can cause health problems, as they can be toxic to humans.

Antibiotics are used so routinely in animals that they may cause some strains of bacteria to become resistant. It is thought they may even leach into cow's milk and may cause sensitivity to antibiotics in humans – one of the reasons why some people are allergic to penicillin.

Some of the hormones fed to animals are used to bring them to maturity quickly. We know that ingesting excess hormones in meat has been implicated in causing health problems, and hormone use has been proven to increase cancer risks in humans – a drug used in the 1960s, diethylstilbestrol, was withdrawn after it was discovered to increase the risk of vaginal cancer for the daughters of women treated with it. And increased amounts of the hormone oestrogen are known to increase the risk of breast cancer.[10] Some hormones can also cause reactions in humans such as unwanted breast growth in men and ovarian cysts.

In the fairly recent past, many naturally herbivorous animals were fed with the waste carcasses and offal of other animals to boost their protein intake and, again, promote growth – a practice that has been implicated in scrapie in sheep and BSE in cattle.[11] The results of this have been well-documented, have been devastating for the UK cattle industry, and the effects are still being felt today.

SUSTAINABLE FISH

The sustainability of fish is a big issue as some areas have been over-fished and stocks are depleted. The result is that traditional fish and chip shops are having trouble supplying the demand for once-plentiful favourites like cod and chips. Here are a few things to consider when buying fish:

- Look for the MSC (The Marine Stewardship Council) logo, which certifies sustainable fisheries. These fish come from healthy stocks and are caught with methods that cause as little harm as possible to the environment.

- Look at labels – the EC now enforces a rule that all fish must be clearly labelled with details of where and how it was caught.

- Some shops only sell fish from sustainable resources (see *Resources*, 9), so give them your custom.

- *Cod*

  Whilst North Sea cod is endangered, cod caught off the coast of Iceland is being fished sustainably.

- *Salmon*

  Farmed salmon is not environmentally friendly as it causes problems like the proliferation of sea lice and pollution. Ideally buy wild salmon.

- *Scallops*

  These can be gathered by divers, a labour-intensive and expensive method, but the alternative is a big net that scrapes them off the sea bed, bringing everything else with it too. So look for the former at the fishmonger's.

What fish should we avoid?

We are told to eat lots of oily fish while pregnant; it is said by some to increase a child's brain power and to help the development of the foetus. This advice has been modified quite severely in recent years, as many of our fish stocks are now contaminated with mercury and other heavy metals, which can damage developing children's brains.

Oily fish such as mackerel, pilchards, herring, trout and sardines are still considered safe to eat and beneficial to your health, as they do not contain harmful levels of dangerous toxins.

The main fish you should avoid are shark, marlin and swordfish, and you should also limit your intake of canned and fresh tuna (bear in mind that canned tuna does not actually count as oily fish as the canning process removes much of the oil). These kinds of fish may contain PCBs (organic compounds which are carcinogenic), dioxins, heavy metals and possibly radioactive materials too. The very oiliness for which these fish are celebrated is the root of the problem – it holds these toxins deep in the flesh of the fish. These contaminants are the by-products of heavy industry and of the incineration of waste which has leached into our water systems. PCBs and dioxins are linked with cancer, disorders of the nervous system and damage to a growing baby, whilst toxic metals like cadmium, mercury, lead and arsenic have been implicated in kidney damage, brain problems and cancer. They are also particularly harmful to unborn children. [12]

Other fish issues

Once you have negotiated the problems of which fish are and are not contaminated, you may then find yourself thinking about the environmental effects of the fishing industry as a whole. It is well-documented that some kinds of fish are being caught almost to the point of extinction and so there are some 'good' and 'bad' choices you can make.

As well as over-fishing and contamination there are other issues which you might want to consider:

The industrial methods of fishing used in our waters, whether trawling, drift-netting or purse-netting, often trap fish, mammals and crustaceans that are not the target of the exercise. Thus dolphins often get caught and drowned when tuna are being caught and turtles can get caught in shrimp and dolphin nets, too. Although since the 1990s, the term 'dolphin-friendly' tuna has been bandied about, more recent fishing methods, which do avoid deliberately catching dolphins, may unfortunately catch sharks, turtles and even seabirds accidentally. So it may be best to avoid that tuna sandwich altogether.

Fish-farming can destroy the plants on the seabed beneath the cages, as the fish are fed a high protein diet which is devastating for the seabed fauna. It can also encourage the proliferation of infection, disease and parasites in the fish, such as sea lice. If you are interested in these issues, Vegetarians International Voice for Animals has more information on how fish are farmed, treated and caught. [13]

A MODERN DILEMMA: GENETICALLY MODIFIED CROPS

What are they?

This is a huge, contentious subject and one that dips in and out of the media on a regular basis with stories of 'Frankenstein Foods' and food to feed all the world's starving millions. [14]

To put it simply, scientists have long been searching for the answer to the problems of food shortages, particularly in the third world. By changing some parts of the DNA of the seed or grain, they have been able to come up with crops that produce high yields and are resistant to whatever problem might be prevalent in the area it is to be grown in – a particular pest, for example, or a vulnerability to drought. They may also be made to be resistant to a particular herbicide, so that crops can be sprayed to kill weeds without damage to the crop. Others produce toxins to kill pests.

Some people argue that farmers have been doing this for centuries by cross-breeding strong strains of plants to get improved varieties. However, genetic modification is a different process from growing better plants by using cross-breeding – it involves changing or adding something to its very DNA make-up. This may be from another plant but it may also be from the DNA of an animal, or even a human.

What's the problem?

One problem is that we simply do not know what the long-term implications might be as this is relatively new technology. There is concern on a number of counts – environmentally, morally and from a safety standpoint:

• Non-GMO crops may become contaminated by the pollen of the modified plants, creating 'super weeds' resistant to herbicides.

• Organic crops can become contaminated by the pollen of GM plants, since it is not possible to create boundaries between crops.

• GMO plants can only be treated with pesticides and fertilisers developed by their creators – essentially making poor farmers in the third world dependent on those companies and their (Western) governments.

• Some people object to changing the structure of plants at all, as it meddles with nature, and the consequences for the environment are unknown.

• Some crops have a so-called 'terminator gene' which means the crops do not produce seeds for replanting the next year. This impacts most on the third world, as farmers cannot save and regrow seed but must buy more – again, making them dependent on the manufacturer.

Though this technology was initially touted as the solution to the world's famine problems, the reality is that the crops currently grown tend to be in the West and are restricted to only a few companies, the best known of which is Monsanto. 75% of the world's GM crops are grown in the USA and Canada.[15] Currently there are no GM crops being grown in the UK and there are no plans for any, but some GM ingredients are present in imported food.

What products are on the market?

At present, GM is mainly restricted to soya, maize, tomatoes and oilseed rape. These are not commercially grown in the UK at present, but may be included in imported products. There have been some GM trials in the UK, but due to consumer pressure, no GM crops are currently being commercially grown.

How can you check to see if a product contains GM ingredients?

Many believe that all GM products or foodstuffs containing GMOs should be clearly labelled as such – including meat from animals that have been fed on GM food. However, current legislation means that the labelling is patchy and whilst some products must be labelled, there is no compulsion to label others. So, in practice, GM products that are sold pre-packaged such as flour, oils and glucose syrups must be labelled on the pack – or if sold loose, a sign should be clearly displayed next to the products. The Food Standards Agency website has the full list of GM products and notes the inconsistencies on labelling.[16]

For animals that have been fed on GM fodder, there is currently no law to induce farmers or processors to label the meat as coming from a GM source, and products made with GM technology (such as cheese made with GM enzymes) does not have to be labelled either.

There is also a fudge about 'intentional' and 'unintentional' inclusion of GMOs – if the product has less than 0.9% GM ingredients, it is considered to be an 'unintentional' use and need not be labelled.

If you are in any doubt, it is safest to stick to organic foods that have the best chance of not containing any GMOs.

Get involved

If you are concerned about genetic modification of our food and if you want to help the fight against GM, write to your local supermarket to ask them to make a commitment not to stock meat or milk that has been produced using GM crops. Several have own-brand meat and milk which is non-GM but are not so choosy about frozen or processed foods. (Marks & Spencer and the Co-op have already changed their suppliers, but others are slow to follow suit.) The Soil Association has information on which of the UK supermarkets are best to choose if you want to buy meat and milk that are GM-free, and they have some shocking findings as to the amount of GM maize and soya being fed to UK farm animals and the health problems this is causing the animals.[17]

If you want more information or would like to get involved in the campaigns against GM crops, there are several UK organisations you can join.[18]

FAIR TRADE

As mentioned in the introduction to this chapter, if you are unable to source local, seasonal and/or organic food, particularly in areas such as exotic fruits and vegetables, the next best thing to do, ethically speaking, is to choose to buy fair-trade produce. This means that if you are unable to source a locally grown, organic product, you can often choose a fair-trade one as the next best ethical (though not necessarily the best environmental) thing. For example, you are now able to choose produce that is not grown in the UK, such as bananas and mangoes, from fair-trade sources and many of them are organic too. These fruits and vegetables will be a welcome addition to your and your baby's diet. However, do bear in mind that any product brought in from overseas incurs food miles, so think about using an alternative, locally available product first.

We've come a long way since the days when coffee and bananas were the only fairly traded product that you could find easily, and today you can make up a larger proportion of your diet from fair-trade products. We still have some way to go with vegetables, though there are quite a few nuts available. Fruit, however, is well represented and you can find apples, avocadoes, bananas, citrus fruit, coconuts, lychees, pears, pineapples and plums, all of which can be included in your diet and your baby's. There are also other ingredients available such as honey, yoghurt and herbs and spices, in addition to products using them as ingredients such as cakes, preserves and confectionery.

In the UK there are rules about what constitutes a fair-trade product, and these are laid down by the organisation that oversees the accreditation of products, the Fairtrade Foundation, which estimates that in the UK alone, sales of fair-trade products are now running at £300 million per year. 2,500 catering and retail items now carry the Fairtrade logo and our supermarkets are working hard to catch up with the phenomenon – some retailers are planning to switch their banana supplies to stock fair-trade only, and others are adding more lines by the day. See *Resources*, 10, for a list of the kinds of fair-trade food you can buy and who stocks it.

The benefits of fair-trade

- The workers are guaranteed a regular price for their products and a minimum price is guaranteed.

- Workers are paid direct rather than being forced to sell their products via dealers and middle men.

- Profits are reinvested into the business or used to provide better conditions for the workers and their families.

- Being paid a fair price means each community can determine how to reinvest profits.

- It allows growers and suppliers to become self-sufficient rather than relying on aid.

- It allows farmers to stay on their land and develop it, also encouraging good farming practices.

- It encourages the growth of co-operatives so the benefits of fair-trade can reach the farmers and their communities.

In addition, many companies who choose to buy fair-trade products also tend to place a premium on getting organic products – and vice versa. That's because people who set up companies to offer ethical choices for consumers tend to be interested in all areas, whether it be fair working practices or growing products which will not cause harm to the planet, animal testing, the use of chemicals and artificial ingredients, etc.

Take action

So if you want to make your shopping pound as ethical as possible, look for fair-trade products in your local shops and supermarkets and, if you can, ask your local shop-keepers to get more fair-trade products. By demanding products that conform to these ideals we are helping to make a difference to the lives of farmers, producers and suppliers directly. This is not a scheme where the results cannot be seen or are difficult to comprehend – there are several personal stories that you can read on the pages of the Fairtrade Foundation (see *Resources*, 10).

Chapter Seven

Inside the House

Every product we use around our houses, on our furniture and upholstery and in the garden will have an impact on the planet, and because we are exposed to these products on a daily basis, either in the air or on our skin, they affect us too. Astoundingly, pollution inside our houses is between two and ten times greater than outside, depending on what decorating products we have used inside your home and what we use to clean it with and make it smell better. [1] The effect of these products on babies, because of their immature systems and thinner skin, can be even greater than on adults, [2] so all products used to clean, polish, descale and remove dirt and mould in kitchens and bathrooms, plus weed-killers, plant feeds and pest-control products in our gardens can be potentially harmful. [3] Take a moment to look at the ingredients on the packets or containers and also consider how they have been made, what they are packaged in, what effect they have on the environment in their manufacture and how they are disposed of.

This chapter gives you a round-up of the natural, chemical-free and organic cleaning and gardening products available. It also tells you the need-to-know facts about many of the mainstream products we use in and around the house, warns of any harmful ingredients they may contain, and describes their effect on us, our babies and our planet.

CLEANING

If you want to keep your home as chemical- and contaminant-free as possible, so as to minimise the effects of these products on your baby and the rest of the family, there are some excellent products on the market for every use around the house. Alternatively, you may even decide not to buy any pre-made cleaning products but to make your own from natural products. Either way experiment with what works best for you and if you don't find one product works, don't give up – try another! (See *Resources,* 1.)

In the kitchen

Washing the dishes

There are now several brands of environmentally friendly, completely biodegradable washing-up products, available in health food shops, some major supermarkets and online; many of these environmentally friendly companies offer washing-up liquids along with soaps and rinse aid for the dishwasher. These products make use of natural ingredients such as citrus oils and non-ionic surfactants. Most brands are fully biodegradable, they tend not to have been tested on animals and they have minimal environmental impact. Many of these washing-up products are packaged in recyclable plastic, so they can be recycled.

> ECO TIP: Buy in bulk to avoid the packaging; use a five-litre bottle, then either use it to refill a one-litre bottle (which is handier for your kitchen worktop), or use a ceramic or glass pump-action hand-soap dispenser. In some places it is possible to take your empty containers and have them refilled, which can be cheaper.

If you do use a mainstream, washing-up liquid, make sure you use as little as possible. Many brands now advise that a small quantity goes further – so take them at their word and just use as little as you can get away with. Soak heavily stained pans to loosen encrusted food before washing, rather than using greater quantities of detergent. Remember that no detergent is designed to be eaten, so you should always rinse your plates thoroughly.

What's the problem with washing-up liquid?

Conventional washing-up liquid contains chemically derived detergent (which is similar to soap but is made from chemical compounds rather than natural fats and lye), which lifts dirt and food remains away from crockery and cutlery.

Chemical detergents can cause pollution to our rivers, streams and seas; when they are poured down the drain they destroy the natural external mucus layer on the skin of fish, which is there to protect them from bacteria and parasites. It can also damage their gills. Another problem is that these detergents lower the surface tension of water, which in turn means that pollution caused by pesticides and other poisons can more easily enter the water and damage aquatic life. When detergent levels reach sufficient quantities, they can kill fish. Much lower levels will kill or damage fish eggs, making it much more difficult for them to breed as usual. Some detergents can also cause some plants and algae to grow faster, which use up oxygen in the water so the fish have less to breathe.

ECO TIP: If you use a dishwasher:

- Always run on a full load to reduce water, energy and soap use
- Measure detergent carefully and don't use too much
- Use an environmentally friendly soap

Some of these detergents also affect humans: washing-up liquids can have ingredients including naphtha, a central nervous system depressant, chlorophenylphenol, which stimulates the metabolism and is toxic, and diethanolamine, which may poison the liver. All these chemicals are washed across your dishes every day and you then eat from them – so there is a good chance that traces will find their way into your baby's system. See the table of harmful chemicals on page 106.

Cleaning surfaces

Many householders choose separate cleaners for their counter-tops and tables. You could actually use one product for all of this, but some feel that surfaces where food is prepared need anti-bacterial cleaners. However, many of the problems associated with washing-up liquids are doubly true of the mainstream anti-bacterial cleaners on the market, which contain chemicals to kill germs and remove mould and fungus.

If you want to use a surface cleaner that has little environmental impact, try one which uses natural ingredients to cut through grease and leaves surfaces clean and hygienic. There are also multi-surface cleaners suitable for all hard surfaces and floor soaps you can use to mop floors. They dry streak-free and have pleasant fragrances.

Grease

If you need to cut through tougher grease, for example in ovens and on cooker hoods, even on barbecues, use a green degreaser.

Blocked sinks and drains

For blocked sinks and smelly drains, try a product that contains micro-organisms that cling to pipes to digest blockages in drains and sewers (see *Resources*, 2). In addition, see the tips on natural cleaning products on pages 102-5.

Around the house

Air fresheners

To keep a room smelling fresh, open a window! Not only does this allow fresh air into a room and smells out, it can also help with keeping moulds, mildew and fungus at bay, which can be a problem in modern centrally heated houses that are often kept almost hermetically sealed, which encourages dampness.

If the air in a room smells stale, there are some great room sprays available, often made by the same manufacturers as the environmentally friendly cleaning products mentioned above. These typically use citrus oils to improve a room's fragrance (see *Resources*, 3). Incense, potpourri and essential oils are other ways to get your rooms smelling nice, and there are fragrances to suit everyone's taste.

Artificial air-'freshener' sprays, or one of the new breed of chemical air fresheners, claim to get rid of smells rather than masking them with fake perfumes, but they are packed full of Volatile Organic Compounds (VOCs) and other poisonous chemicals (see Chapter 4, page 64). The propellants (the chemicals in the bottle that allow the spray's ingredients to be carried on a fine mist) can cause nerve damage to both adults and babies and in addition are inflammable – and of course, artificial perfumes can cause allergic reactions.

Limescale

A perpetual problem in hard water areas, many of us feel we have to reach for heavy duty limescale removers to keep our appliances, toilets and kettles gleaming. However, there are other ways and some of them quite ingenious – read on! (see *Resources*, 4)

There are a few environmentally friendly limescale products, suited for cleaning small amounts of scale from around taps, on baths and sinks. However, as with all things, prevention is better than cure and magnetic devices seem to work best, though it's quite a mystery how they do so; put simply, the magnetic field keeps calcium molecules apart and reduces the amount of particles that are able to 'stick' to the insides of the pipes.

There are several devices available that you can put into the washing-machine or into the cistern of the toilet. Over time, they prevent the build-up of scale with no chemicals and no need to do anything further. There is also a similar magnet that prevents all your water pipes from clogging up with limescale. It fixes onto the main cold water supply pipe coming into your home, so that all your water pipes throughout the house benefit. This will enable you to cut out the calcium-reducing tablets you put in the washing-machine – after all, if the tablet can dissolve limescale, what on earth is it doing to your clothes and skin?

> ECO TIP: Beeswax is a good, natural alternative for wood as it actually nourishes it and keeps the furniture clean and shining. Experiment with other natural substances like lemon oil, which not only gives a wonderful shine to light wooden furniture but smells fantastic, too. Wooden instrument makers often use it to clean the fingerboards of guitars and violins. Even olive oil can be used on wood, but buy it from a pharmacy as it will be low in fragrance. You don't want the wardrobe smelling like a restaurant!

Polish

If you just want to remove dust, a slightly damp cloth works just as well as a furniture polish spray – perhaps better, as it simply removes the dust, yet leaves no residue. Spray polishes are unsuitable for some surfaces, such as wood that has been French-polished.

The propellants and artificial perfumes in furniture polish can cause irritation to the eyes, skin and lungs and some even contain formaldehyde, which may cause cancer. On a practical note, they can cause a build-up of product on your furniture, which in time dulls the surface of your furniture and can crack.

A wonder cloth

If you want to cut out several cleaning products all together, get hold of an E-Cloth. It is made of millions of tiny fibres that have a good, natural cleaning effect and the cloth is suitable to clean absolutely everywhere around the house. Use dry or add just a touch of water (see *Resources*, 5).

Doing the laundry

Instead of chemical cleaners, look first for biodegradable laundry products that are derived from plants and that contain no artificial perfumes – there are plenty that contain natural ingredients like lavender and lemon! (see *Resources*, 1).

When washing nappies, make sure you take advantage of the new generation of nappy sanitisers (see Chapter 3).

Mainstream laundry soaps and fabric conditioners are a major cause for concern if you want to keep chemicals and artificial substances away from your baby's delicate skin and sensitive respiratory system – particularly important when the fabrics washed will be placed next to your baby's skin which is five times thinner than yours, thereby allowing more toxins to pass through into the baby's system.

Some laundry products also contain the same petrochemically derived detergents as mainstream washing-up and cleaning products, plus more fragrances, with all their associated problems. In addition, they contain chemical substances called optical whiteners, designed to make your white clothes look whiter but which can cause skin irritation.

What – no soap?

Alternatively, try one of the no-soap cleaning methods. The sceptics among you will wonder how on earth these things work if there is no soap or detergent involved – but rest assured that I have been washing my family's clothes with these ingenious devices for years, and they work!

Eco Balls or Aqua Balls are plastic capsules, inside of which are little pellets. As they move around inside the washing machine, they react with the water to release ionised oxygen, which gently yet effectively lifts the dirt from the fabric (see *Resources*, 6).

There are several advantages:

- They are really cheap to use as, although they cost quite a lot to buy initially, each wash costs around 3p and they last up to six years, saving you up to £200 on detergents.[4]

- There's no measuring of liquid or powder.

- You can wash at a low temperature – as low as 30ºC, which in turn saves energy (and more money).

- You can cut out the extra rinse cycle if your brand of washing-machine offers this option.

In the bedroom

Baby's mattress

Organic mattresses usually have an unbleached cotton cover, fabric which has been grown without pesticides and which has not been chemically treated. It is generally removable and can be machine-washed (if there is only light soiling you can do this at 30ºC, though it is recommended to wash every now and again at 60°C, which is the required temperature to kill bacteria and dust mites).

When you clean the mattress, you should use products that will not have a bad effect on your baby – products that are best for baby's skin and will not give off

NATURAL TIPS TO KEEP MITES AT BAY

- Enclose mattresses in a dust-mite-proof cover, available from many allergy websites and stores. Fitted sheets can have a similar, though less effective result. On a cot, use only a cover designed for cots, not a plastic cover which can cause overheating. Do not use pillows with a baby under one year.

- Regularly vacuum the mattress and pillows, and the base, sides and headboard (if upholstered) of the bed.

- Wash bedding regularly (some suggest every two weeks for a severe allergy), at 60ºC. Freezing also kills them, so put blankets with low recommended wash temperatures, cushions and pillows in the freezer every now and again.

- Toys can also harbour mites, so go for washable soft toys or wooden ones.

- Mites love warm, damp conditions so keep heating low, the room well ventilated and air the bed well each day. If the room is very damp, consider buying a dehumidifier.

- When you clean, use a damp cloth or brush to gather dust and mites rather than stirring them up. Vacuum the carpet daily for severe allergies.

- Avoid carpet and too much fabric in the room. A wooden floor is much easier to damp-clean and harbours less dust – therefore, fewer mites!

harmful chemicals, like those mentioned above. Also, be very careful of cleaning a plastic-covered mattress with anything chemical – it may cause the plastic to deteriorate or give off fumes. Use an environmentally friendly fabric cleaner or mild soap flakes.

Dust

Dusting is one of those chores we tend to put off until we can actually write our name in the accumulation on the shelves! The best, most environmentally friendly (and cheapest) way to dust is with a damp cloth. This is generally enough to remove dust as it traps it on the cloth. Flicking dust around with a feather duster raises it and just redistributes it around the room! If you have an asthma sufferer in the house, the damp cloth method is probably the best, as it removes dust and does not introduce any allergens.

After dusting, avoid using commercial products to polish with – you might get a gleaming piece of furniture and the aroma of country pines, but you will also be spraying chemicals and artificial perfumes around the house. Try some old-fashioned products – for wooden furniture, use humble old beeswax, available in most hardware and furnishing stores – though avoid any with coloured additives.

Dust mites

Dust mites are minute creatures that live on dead skin cells. You can get 100,000 of the little beasts in just one square yard of carpet – and yet they are too small to see with the naked eye. Their droppings contain proteins, which many people are allergic to and they can aggravate asthma and allergies. They tend to live in areas where we lounge about and shed the most skin cells – that is, in our beds, sofas and chairs and the carpets surrounding them.

Don't be tempted to use a chemical carpet cleaner to get rid of dust mites in carpets. Research shows that natural methods actually work better and do not introduce allergens. Regular vacuuming can help reduce their spread. If you feel you need a little more help and really can't get rid of or clean your carpets effectively, you can try a spray with natural ingredients which is deadly to dust mites but completely harmless to humans and animals. Over a period of weeks the dust mite population decreases. There is also a fabric cleanser on the market which you add to the washing-machine that kills mites even at low temperatures (see *Resources*, 7).

In the bathroom

This room is often home to some of the most dangerous chemicals around, as products designed to clean or unblock the toilet and pipes may contain sodium hydroxide and sodium hypochlorite (bleach) that can burn the skin and eyes quite severely, sometimes resulting in permanent damage. They can kill if swallowed, and so it is of crucial importance to keep them well out of the way of an inquisitive baby.

You can use some of the environmentally friendly cleaning products I have already mentioned to clean your bathroom, and experiment with some of the gentle, biodegradable products to remove limescale; consider fitting a magnetic limescale remover on your pipes and invest in one for your toilet, too.

The toilet

There are several environmentally friendly toilet cleaners that do not rely on bleach and yet work just as efficiently. If blockages are a problem, rather than using a chemical unblocker as a first resort, try to get rid of the blockage first!

It is also largely unnecessary to use toilet blocks and hanging perfumed devices to give the toilet a pleasant smell – these are formed from artificial fragrances, detergent and bleaches and can cause all of the problems described above. Keep the bathroom well aired and the pipes clear and smells should not cause a problem.

Sinks and baths

Best for baths and sinks are mild cream cleansers, which lift dirt off without scratching the enamel or porcelain. For blocked pipes use a sink plunger, or a pipe-unblocking tool if the problem is lower down, flush well with hot water and try not to let mess get into the plug hole in the first place – the worst offenders for blocking the plug hole are a combination of hair and soap, which coagulates and sticks in the pipe. You can use a perforated metal or plastic disc over the plug hole, especially in the shower, to catch debris.

Some additional benefits of green cleaning

When you buy environmentally friendly cleaning products you will often find extra benefits on the label, for example many are:

- Perfume-free.

- Free from artificial ingredients which might dry or harm the skin and your health.

- Not tested on animals.

- Packaged in recyclable containers.

- Available as refills or in large containers, thus cutting down on packaging.

Tips for using eco-friendly cleaning products

Most environmentally friendly products will not bubble up as much as conventional ones, as they lack the highly foaming detergents that are contained in mainstream products. However, just because you don't see bubbles, don't think it won't work and don't be tempted to add more.

Also, you may need to work slightly harder at cleaning than with a chemical cleaner, which will simply strip off grease and dirt using the power of its detergents. But think of the extra benefits – more time spent on scrubbing the sink means fewer hours in the gym to get rid of those loose bits of underarm flesh – not to mention a cleaner, safer home for your baby, no nasty residues to irritate his lungs and skin – and a cleaner planet!

The following environmentally friendly cleaning products are also available:

- Glass and mirror cleaners.

- Cream cleaners suitable for enamel and porcelain.

- Limescale removers.

- Dishwasher powder, liquid and rinse aid.

- Polish suitable for wooden furniture.

- Stain removers.

Make your own green cleaners

Our grandmothers had many simple ways of keeping house in the past that were probably every bit as efficacious as modern products – and most of them were far kinder to the environment and our bodies, too. Here are some old-fashioned tips for cleaning – the natural way, mostly without chemicals, and with easily available household products:

Baking soda

- use a paste made with baking soda and water to clean ovens – coat them with the paste, leave overnight, then scrub off using gloves.

- To clean silver, line a heat-proof bowl with aluminium foil, pour in boiling water and half a cup of baking soda. Then place the silver items into the bubbling liquid. Leave in for a couple of minutes then remove and rinse.

- An open box of baking soda gets rid of smells in a room or the fridge. Keep away from children.

- A thick paste of baking soda and water cleans sinks, baths and work surfaces.

- A quarter cup added to your washing-machine will soften clothing.

- Pour onto food stains on the carpet and vacuum up the next day. Also gets rid of pet smells.

Borax

- A mixture of borax with lemon juice, washing soda or white vinegar will remove stains from clothing.

- Use to clean dishwashers.

- A paste of borax, salt and vinegar will clean stubborn carpet stains. Put on the spot and vacuum up the next morning.

Cedar wood

Blocks of cedar can be used to add a subtle fragrance to a room and to prevent infestations of clothes moths. You can buy cedar hangers and rings of the wood to slip in your drawers and cupboards.

Flowers and herbs

Dried flowers such as lavender, lemon verbena and others can be used to give a room a pleasant smell. You can make your own potpourri from your garden or buy some – but avoid any with artificial perfume. Freshen it up every now and again with a drop of essential oil. Lavender keeps insects and moths at bay.

Hot, soapy water

This kills food-related bugs such as salmonella and E coli. Use to cleanse boards, knives, and surfaces that come into contact with raw meat or eggs.

Lemon juice

Use to polish copper, with a little salt dissolved in it.

Lemon oil

Also good as a furniture polish and it smells great, too. It can also be used to deter cats from messing and scratching in specific areas – place a few drops onto the area.

Olive oil

Polish furniture with a mix of one teaspoon of olive oil mixed with half a cup of white vinegar.

Toothpaste

Use to clean silver: smear on, rub in and remove with a soft cloth.

Washing soda

A paste mixed with water is good for more stubborn stains on surfaces. Wear gloves as it can sting a little.
Pour a quarter of a cup of this down the drain each week, followed by hot water, to prevent clogging.

White vinegar

Use to wipe over surfaces to kill bacteria, mould and viruses, it has many uses:

- Dissolve a small amount of salt in it and use to polish copper.

- A paste made with a cup of vinegar, a teaspoon of salt and a cup of flour mixed together can clean brass.

- Fill a spray bottle with water and a cup of vinegar to wash windows.

- Use old newspapers to get a shine.

- Use a quarter of a cup in the washing-machine to soften clothes and to eliminate cling.

- Use for laundry stains as a spot cleaner.

- Use equal parts of vinegar and water to clean stains on carpets. Spray on and then mop up after a few minutes.

Troublesome stains

Mud on carpets
Rub salt on to the mark, leave for an hour and then vacuum.

Coffee
Rub the stain with soda water, then sponge up.

Red wine
Use white wine on top, then sponge up. Alternatively, pour a generous amount of salt on while wet, then brush or vacuum the next day.

Chocolate
Make a paste of Borax and water, rub onto the stain.

Grease and oil
Pour cornmeal or cornstarch onto the mark, rub in, then vacuum up later.

Chewing-gum
Put the affected item in the freezer if it is small enough; otherwise, rub ice cubes on it until frozen, then pick off.

Why avoid conventional cleaners?

We are led to believe that the vast array of cleaning products for cleaning sinks, dishes, floors, ovens, window, mirrors, and toilets, as well as the bleaches and descalers on supermarket shelves, are essential if we are to keep dirt, grime and especially bacteria and germs at bay. TV adverts show us how these mainstream products cut though grease, kill germs and keep our homes limescale-free and sparkling. What they do not show us is a list of their chemical and artificial ingredients – or what harm they can cause, not just to the environment but to our skin and our health, too.

Many of the products we keep for cleaning contain poisonous substances and if they are not properly and safely stored, children are at risk of spilling these and being exposed to their fumes. Worse, they might accidentally drink them.[5] It is thought by some that the fumes some products give off, even through their packaging, account for headaches and depression in adults and an increase in ear infections and diarrhoea in infants.

Most cleaners are derived from petroleum, a fast-diminishing resource and some also contain substances called alkylphenol ethoxylates which may disrupt hormones in humans, are not biodegradable and can adversely affect plants and animals. Some alcohols used in detergents are suspected of causing cancer.

Conventional cleaners also contain artificial perfumes, which though they give a pleasant fragrance, contain phthalates, chemicals derived from plastics to prolong the scent. These are also linked to some forms of cancer and in laboratory animals have been shown to cause adverse effects to their reproductive systems – so by extension, they may have the same effects on humans, especially in babies developing in the uterus. They can also cause asthma, allergies, headaches and sore eyes.

Some ingredients are fairly safe on their own but dangerous when combined with others – such as some of the commonly used preservatives which can react with nitrates to form cancer-causing substances called nitrosamines, as well as chlorine and ammonia, which together give off a poisonous gas.

HARMFUL CHEMICALS FOUND IN THE HOME [6]

Ingredient	Found In	Side Effects
Alcohol	Many cleaning products	Skin dryness
Alkylphenol ethoxylates (APEs)	Detergents	Disruption of hormone function in animals and possibly humans
Ammonia	Many cleaning products, fertiliser	Can cause breathing problems, especially for children with asthma
Benzene	Many cleaning products, fertiliser	Toxic to inhale or ingest, carcinogen
Bleach	Toilet and bathroom cleaners	Fumes can burn eyes and lungs
Creosol	Disinfectants, wood preservers, fertilizers	Respiratory tract irritation
Chlorophenylphenol	Detergents	Can cause unwanted metabolic changes; toxic
Diethanolamine	Detergents	May poison the liver
Formaldehyde	Many cleaning products	Considered by some to be carcinogenic
Paradichlorobenzenes	Toilet cleaner, room deodoriser	Burns eyes, skin and lungs
Phosphates	Dishwasher detergents	Damage to river and marine life
Perchloroethylene	Dry-cleaning fluids, de-greasing cleaners	Nerve, liver and kidney damage, potential carcinogen
Polychlorinated biphenyls (PCBs)	Furniture polish	Irritation of eyes, nose, mouth, damage to foetus, highly polluting
Lye (sodium hydroxide)	Household cleaners	Can burn tissues, especially the eyes
Naphtha	Detergents	Central nervous system depressant
Naphthalene	Moth balls	Damages red blood cells
Optical whiteners	Laundry detergent	Can cause skin irritation
Sulphuric acid	In aerosol products	Irritation of skin, eyes, nose and mouth, carcinogenic
TCE (trichloroethylene)	Paint, paint stripper, adhesive, de-greasing products, dry-cleaning fluid	Irritation of throat and nose, depression of central nervous system, more serious symptoms on high exposure

> ### SAFETY TIP
>
> Keep any cleaning products in your house secure, away from babies and small children, in a kitchen cupboard with a childproof lock. All bottles should be kept sealed tight.

The contents of some conventional cleaners

Opposite is a list of just some of the chemicals that are found, in varying amounts, in commonly used household cleaners. Steer clear of products containing these and look for natural alternatives instead.

IN A NUTSHELL: HOW TO BE A GREENER HOUSEHOLD

If you want your baby to grow up in as green a household as possible, and if you are determined to make a difference to the planet's future, here are some quick tips for changes both big and small that you can make around the house. In addition, make use of the many energy-saving gadgets now available to help you cut down on your energy use and carbon emissions. These include low-energy light bulbs, meters to tell you how efficient your electrical appliances are, and devices to save energy when your computer is on standby (see *Resources*, 8).

Save energy

- Change all your light bulbs to energy-efficient varieties

- Cook wisely: use the oven for several dishes, use the correct size ring for your pan and cook more than one thing in a pan. Only boil enough water for the hot drinks you need.

- Get greener appliances – look for A-rated models where available and reduce your energy use by up to 40%.

- Try to avoid using a tumble-dryer.

- Turn off TVs, computers, washing-machines and stereos – anything with a display – rather than leaving them on standby.

- Turn off all lights not being used.

- Change to a condensing boiler to heat your water.

- If you have a hot-water tank, insulate it well, as well as all pipes.

- Insulate your loft and cavity walls.

- Turn your heating down by 1°C – if you are cold, put on a jumper.

- Change your energy supplier to one sourcing it from renewable energy.

Save water

- Have a shower, not a bath – or share the water.
- Turn the tap off when brushing your teeth.
- Fit a cistern block in the loo to reduce water use.
- Reuse bath and washing water on the flowers.
- Fit water butts on down-pipes to collect rainwater.
- Fix any dripping taps.

Reduce, reuse, recycle

By reducing the amount of 'stuff' you buy, you can actually cut down on the amount you need to throw away or recycle. Here are some tips:

Reduce your waste

- Buy a reusable shopping bag or a bag for life.
- Don't buy packaged fruit and veg – get it loose from the greengrocers.
- Avoid using a separate bag for each kind of fruit or veg – put it all loose in your shopper.
- Avoid over-packaged products; join the growing army of disgruntled consumers who take the packaging off at the supermarket checkout, leaving it for the shop to deal with.
- Buy in bulk where possible to reduce packaging.
- Put left-over food into smaller, lidded bowls rather than using cling film or foil.

Reuse where possible

Get milk delivered by the milkman and return the bottles to be reused.

If you have plastic shopping bags be sure to reuse them. (Some supermarket delivery schemes will take back bags and give you 'points' on loyalty cards for them.) Buy products in refillable containers.

Recycle

There are a few gadgets and devices that can help you – a can-crusher will reduce the amount of space taken up in your box, bins with different compartments help you sort waste from recycling.

- Recycle vegetable and fruit scraps in the compost bin if you have one.
- Pass on your used baby clothes.
- Give unwanted household items to charity shops.
- Take your bottles to the bottle bank if you don't have a doorstep collection.

- Reuse envelopes using charity labels.

- Recycle paper, cardboard and newspaper.

If one person recycles a piece of paper it doesn't make much impact, but if every single person in this country recycles one piece of paper – that's 60 million pieces of paper! (see *Resources*, 9).

Chapter Eight

Toiletries

CHANGING TO MORE NATURAL PRODUCTS DURING PREGNANCY – AND BEYOND

When you discover you are pregnant you may want to think about changing your toiletries for more natural alternatives; chemicals from hair dyes, nail varnish, deodorants, cosmetics and perfumes can be absorbed into the skin via your bloodstream and internal organs and during pregnancy they can cross the placenta into your baby's developing system.[1] The potentially harmful ingredients and their effects are listed on Page 117, but the fact that potentially cancer-causing preservatives have been found in breast milk means it is wise to read the labels of the products you have in your bedroom and bathroom and to investigate all the wonderful, natural alternatives. There are lots of lovely organic cosmetics, creams, hair preparations, soaps and body washes that are based on natural plants, oils and flowers and that smell great, are good for you and your baby – and that work!

'NATURAL' AND 'BOTANICAL' PRODUCTS: MAKE SURE YOU READ THE LABEL!

Beware of mainstream, chemically derived products which claim to be 'natural', 'botanical' or have a similar description – some are just the same old products, packed with artificial ingredients, which have had an injection of plant-based oils or fragrances.

NATURAL COSMETICS

Look for ranges of cosmetics that rely on natural minerals instead of artificial colours. Most organic make-up contains no parabens, mineral oil, synthetic colours or fragrance and most are suitable for vegetarians, vegans, those with sensitive skins and those allergic to mainstream products. There are some great products for the eyes, lips and face and you can even find natural alternatives to nail varnish.

The additional benefit of these products is that they often contain ingredients that soothe and nourish the skin; so, for example, lipsticks may contain jojoba to moisturise, face powder may contain calendula and Vitamin E, and foundation may have extracts of aloe vera.

NATURAL TOILETRIES FOR YOU

Deodorant

You may be reluctant to try one of the alternative deodorants, as we seem to feel it is taboo to have any form of body smell – or even sweat, which is the body's perfectly natural way of cooling down and excreting toxins. But it is worth trying out some of the products available (see *Resources*, 1), either a liquid spray, a roll-on, or a stick deodorant. You could also experiment with natural crystal deodorants, which derive from saltrock or ammonium alum sulphate. You may find it takes a little while to get used to any of these, but you'll find that rather than smelling bad, you will smell different: the products don't block up your pores, but rely on natural ingredients to neutralise bacteria that make sweat smell.

Essential oils

Always use any oils with caution during pregnancy and ask your health professionals and the retailer for advice on what is and is not suitable for use whilst pregnant, when breast-feeding or on a baby (see *Resources*, 2). Bear in mind that lavender and tea tree oil should not be used on babies or when you are pregnant.

Hair dye

Avoid using dyes, and especially bleach, on your hair whilst you are pregnant or breast-feeding and particularly avoid any lightening products. If your hair needs a little help, try a herbal or semi-permanent hair colour like henna, which is a good, natural product.

Make-up remover and cleanser

Try to avoid the commercial cleansers and alcohol-based toners, all of which can be very drying and remove make-up with a natural product like almond oil or an organic face wash. Calendula, chamomile and mallow all replenish the skin as they clean and are soothing.

Perfume

You may not realise it but many mainstream perfumes contain plasticisers called phthalates which are designed to keep the fragrance on your skin for longer, thus prolonging the scent. They will also contain preservatives. As both these are potentially harmful to your and your baby's health, why not simply avoid using perfume, especially while pregnant and during the breast-feeding months.

Shampoo

Change to natural, organic shampoos and conditioners, thus avoiding all those harsh, drying chemicals and reap the benefits of the added boosts of organic ingredients like rosemary, lavender, calendula, hemp and aloe vera.

Soap and body wash

Avoid harsh, drying body washes and soaps and switch to natural soap products such as olive oil soap produced by the Aseela Women's Co-operative in Bethlehem (see *Resources*, 3) and help support disadvantaged women there at the same time. Alternatively, look for a shower gel or body wash infused with flower fragrances.

Toothpaste

Pregnancy can be tough on the teeth and gums, with the hormone changes caused by the condition often resulting in tender, bleeding gums and even tooth loss if you're not careful – the old wives' tale, "lose a tooth for each child", is not so far-fetched, based on the effects that our hormonal changes have on our dental health. Added to this is the fact that the major detergent found in toothpaste, sodium lauryl sulphate, can cause skin cells to shed. A study in Norway in 1996 compared the use of toothpastes that contained this detergent with those that did not, and found that 21 out of 27 participants experienced loss of the cells on the gums when using the SLS toothpastes.[2] Always read the labels, as some toothpastes sold in health shops do have unwanted chemicals in them but are sold for other beneficial effects like their whitening power. There are plenty of natural alternatives. Once you get used to the difference in texture – and the fact that natural toothpastes do not bubble in the mouth – you'll never go back to mainstream toothpaste; who wants a mouthful of bubbles that are caused by detergent?

Find out more

To find out more about the hidden chemicals in our everyday cosmetic products, see the studies and research on the Women's Environmental Network website,[3] or see research by Greenpeace,[4] in which, during a study of blood taken from umbilical cords in the Netherlands, scientists found the following chemicals: brominated flame-retardant TBBP-A, phthalates, artificial musks, bisphenol-A, triclosan, alkylphenols, organochlorine pesticides (DDT) and perfluorinated compounds.

NATURAL TOILETRIES FOR YOUR BABY

If you want to ensure you keep your baby away from chemicals and ingredients that might affect her health, the best thing is to choose products marked 'pure' or, ideally, 'organic'. Read the ingredients of products marked 'natural' or those that claim to be 'plant-derived' to make sure they are not just mainstream products masquerading as natural ones. Make sure you check out the criteria the company uses in sourcing its products – the Soil Association logo shows you that a product is certified organic. Organic products are subject to stringent requirements and will therefore be purer – but may have a price hike to compensate! However, a little goes a long way and the price will reflect the pure, natural ingredients found in the products and the way in which those ingredients have been grown and harvested. Some products, for example, rely on botanical ingredients like calendula and roses that have been grown organically and use natural preservatives (see *Resources,* 4).

BABY SKIN PROBLEMS
AND NATURAL SOLUTIONS

No matter how careful you are with your baby's skin, there will always be times when she has a little nappy rash, dry skin or minor problems like insect bites and stings. Some children are also prone to more serious skin conditions like psoriasis and eczema. Whatever the problem, there are a whole host of beneficial plants and soothing ingredients that are available in organic preparations. Here are some of the most common problems and solutions that are good for you, your baby and the planet.

Dry skin

Most of us have dry skin from time to time and the key here is not to try to put something on to seal it first – we should get water into it initially, then use a barrier or moisturising cream on top. So a good soak in the bath will hydrate baby's skin before you apply a cream on top. Either choose a moisturiser with ingredients like olive oil, or use something with a beeswax base to seal the moisture into the skin. Avoid soap, baby bath and body washes.

For weather protection use creams containing beeswax and calendula which can help keep baby cheeks and hands soft and supple in cold weather. And between baths, use mild, organic creams, lotions and salves to keep the skin soft, soothed and moisturised.

Eczema

You may find that some things work for your baby and others don't, so be prepared to try different products. Conventional medicine will often only offer you steroid creams, which can thin the skin, or emollient creams and lotions, which may help by sealing in moisture but often contain artificial preservatives and alcohol, which is drying. Keeping the skin cool helps a great deal, as does using organic clothing, plus avoiding soaps and body washes which may be drying.

As eczema may have diet triggers, discuss this with a nutritionist or an allergy expert (many pharmacies now have them as a regular feature), or try eliminating certain weaning foods for a while to see if it improves – tomatoes are often culprits. You may need to change your diet if you are breast-feeding – eczema can be exacerbated by what you are eating, for example dairy foods and citrus fruits.

As for dry skin, the best tip is to hydrate the skin as much as you can – and that means long, long bath times! After bathing, the choice of cream, lotion or oil you use to seal in the moisture depends on what you find works best for your baby – a certain amount of trial and error may be needed. Generally, creams containing a mixture of ingredients like calendula and red clover have good results, and can help address that side of the condition. There are also some preparations with royal jelly, propolis and aloe vera which worked well for my son.

Nappy rash

Most babies are prone to nappy rash at one time or another during their young lives. It may be when a new weaning food is introduced – tomatoes are prime offenders. It may be because of an allergy. When teething, your baby will be prone to producing extra acid in his saliva, which causes a sore bottom! Nappy rash can also be caused by gels in disposable nappies; if so, try changing brands to see if it makes a difference.

The best advice is to try to prevent nappy rash as much as you possibly can and this means changing the nappy as soon as it is wet, cleaning urine and faeces from the skin and then using some form of moisturising or barrier cream to prevent chafing. There are several nappy balms which use plant-based ingredients to keep the skin healthy, but ideally use water and mild soap if skin is soiled rather than baby wipes. Letting the skin get plenty of air helps too, so if you can, leave your baby without his nappy on for a while.

The best nappy rash cream I have ever come across was an organic calendula nappy cream. My son often got a sore bottom but if I applied this cream at night, it was almost always clear by the morning. We called it 'magic cream', and there is always a tube in my bathroom cabinet. Clover creams will also do a good job.

Small cuts and bruises

For cuts, clean the area thoroughly before applying a product such as a spray or cream with a mixture of natural antiseptics mixed with soothing ingredients like aloe vera and restorative ones like calendula.

If your baby has bumped himself, applying arnica can help prevent bad bruising. If the bruise appears without you having noticed a bump, it's not too late to apply this remedy, as it will stop the bruise worsening, but use with caution: bruising can indicate an underlying problem, so if you are unsure, see a doctor. If your baby has to have any kind of operation or hospital procedure, smoothing arnica gel on to the area for a few days before will help keep bruising and swelling to a minimum.

THE HIDDEN DANGERS IN TOILETRIES

In the UK we spend £5 billion each year on cosmetics and toiletries.[5] About 93% of us use cosmetics in some form, meaning that British women are among the highest users in the world.[6]

We are exposed to chemicals in several ways; our skin, the body's biggest organ, absorbs them, we wash them down the drain when we bathe, and they then return in our drinking water, and we unwittingly consume them.

Although potentially harmful chemicals contained within cosmetic products are in minute amounts, it is our constant and prolonged exposure to them that causes concern. Some chemicals can cause allergies, some are known to disrupt hormones and others have been linked to rising incidences of birth defects, infertility in men and early puberty in girls. The irony is that products targeted at making us more virile, look younger and become fitter contain ingredients which can cause premature ageing, disrupt hormones and can cause health problems, including cancer.

In a random study of cosmetics (including make-up, hair and body-care products), the Women's Environmental Network found preservatives suspected of mimicking oestrogen in an astounding 57% of products – which is worrying when you consider that an unnatural increase in oestrogen can cause breast cancer.[7] This has also been implicated in a drop in male sperm counts and to an increase in breast and testicular cancers. Whilst some scientists believe the link is a great threat to public health, others remain unconvinced of the cause and effect of the preservatives contained in our cosmetics.

Products such as bubble baths may also contain a probable human carcinogen known to cause cancer in animals called 1.4-Dioxane. As this is produced during manufacture – i.e. it is a by-product of the manufacturing process, not an added ingredient – it is often not mentioned on the ingredients list.

How do you make informed choices about the toiletries you use when you are pregnant, and the products that you use on your baby? When you consider that some wetting agents and solvents in make-up, hair products, deodorants and after-shave use the same ingredients as those used in antifreeze and brake fluid, that detergents are present in our children's toothpaste, that cancer-causing chemicals are in our perfumes and that an agent found in skin cleansers is also used to dissolve grease, you might think again about choosing mainstream products.

Controversy over talcum powder has been discussed in the medical press for decades, and mothers are advised not to use talc on babies, especially girls. Therefore it is naturally confusing that some baby products still contain talc-based ingredients.

INGREDIENTS TO AVOID

Here's a run-down of the most commonly found ingredients, what they're for and what effects they have on your health and that of your baby. Take a look at the products in your cupboards and on the shelves of pharmacies and supermarkets and see how many of these are listed on items you use every day.

Name	What is it?	What's the problem?
DEA (diethanolamine), or TEA (triethanolamine) Found in soaps and shampoos	Petrochemical detergents or boosters, can be used as thickeners	Have been known to cause allergies, irritate eyes, drying to hair and skin. Potentially cancer-causing, creating problems in kidneys and liver
Imidazolidinyl urea, DMDM hydantoin Found in body and hair products, antiperspirants and nail polish	Preservatives that form formaldehyde	Linked with headaches, depression, dizziness, allergies, asthma, joint and chest pain, chronic fatigue syndrome and insomnia. May weaken the immune system, may cause cancer
Isopropyl Found in many products including hair colour, hand and after-shave lotions and fragrances.	A form of alcohol	A poisonous solvent and it has a denaturant effect, changing the structure of other chemicals. It has been known to cause nausea, vomiting, headaches, hot flushes and depression. Can dry skin and hair, and encourage growth of bacteria
Petrolatum Found in baby oils and jellies	Mineral oil jelly, also known as baby oil, vaseline, *paraffinum liquidum*	Can take natural oils from the skin leaving it dry, causes chapping in nappy area, also premature skin ageing. Forms a barrier stopping toxins being released from the skin. Can cause acne
Padimate-O, octyl dimethyl or PABA Found in sunscreens	A nitrosamine-forming agent like DEA	It is thought the sun may react with it to form free radicals, which may increase the risk of skin cancer
PVP/VA copolymer Found in hair spray	A petroleum derivative	Can cause dry skin and itchy scalp
Sodium lauryl sulphate (SLS) or sodium laureth sulphate Found in many products including conditioner, shampoo, cleansing washes, toothpaste	A detergent	Can affect eyes, causing irritation and even permanent damage, especially in children. Has also been implicated in rashes and flaky skin, hair loss, mouth ulcers. If mixed with other ingredients, can form nitrosamines, which are carcinogenic. It soaks into skin very easily and can be stored in major organs
Methyl, propyl, butyl and ethyl Parabens	Preservatives used to extend shelf life of products and prevent growth of microbes	Highly toxic, can cause rashes and allergic reactions.
Synthetic fragrances Found in many cleansers and cosmetics	Petroleum-based	Can cause headaches, dizziness, rashes, respiratory problems, vomiting, skin irritation and sensitivity to other chemicals
Talc or talcum powder Found in baby, face and body powders, on contraceptives and medical gloves	A powdered stone	A known carcinogen, has been implicated in causing ovarian cancer when used in the genital area. Can lodge in lungs and cause respiratory disorders
FD7C or D4C followed by a number Found in many cleansers and cosmetics	Synthetic colours, derived from coal-tar	Carcinogenic

GET INVOLVED

If you want to find out more about potentially harmful ingredients in our cosmetics and toiletries, see the websites of The Women's Environmental Network,[8] the Campaign for Safe Cosmetics,[9] which campaigns for safer and more natural ingredients in our toiletries, and organisations like Greenpeace,[10] which has some interesting research studies. There is also information available on the potentially dangerous phthalates contained in baby lotions, shampoos and washes.[11]

Notes

Chapter 1: Getting Ready for your Eco Baby

1. Child Accident Prevention Trust. www.capt.org.uk. Tel 020 7608 3828.

2. Baby Products Association. www.b-p-a.org. Tel 0845 456 9570.

3. Environment California Research & Policy Centre report: *Toxic Baby Bottles: Scientific Study Finds Leaching Chemicals in Clear Plastic Baby Bottles*. The report describes the harmful effects of the hormone-disrupting chemical bisphenol-A (BPA), a developmental, reproductive and neural toxicant found in polycarbonate plastic – the material used to make the vast majority of baby bottles. February 2007.

4. See 2 above.

5. Foundation for the Study of Infant Deaths. www.fsid.org.uk. Helpline 020 7233 2090.

6. *Reviews of Environmental Contamination and Toxicology* by George Whitaker Ware, Vol 184, Feb 2005, published by Springer (www.springer.com), and 'PBDEs in breast milk: levels higher in United States than in Europe', Science Selections from *Environmental Health Perspectives*, Nov. 2003 by Charles W. Schmidt, National Institute of Environmental Health. Scienced-based research group Sightline (www.sightline.org/research/pollution/res_pubs/health_concerns).

7. Women's Environmental Network's Breasts and Bellies campaign, 15th Jan 2002. (www.wen.org.uk/general_pages/Newsitems/pr_breastsbellies.htm).

8. News item on sheep-dipping reported by BBC. news.bbc.co.uk/1/hi/health/383003.stm. Also: Institute of Occupational Medicine (www.iom-world.org/news_archive/rethink.php): 'Farmers Demand Government re-think on Organophosphates Guidelines: Epidemiological research at IOM', led by Dr Brian Miller.

9. Department for Environment, Food and Rural Affairs' advice on removing lead paint. www.defra.gov.uk/environment/chemicals/lead/advice3.htm.

10. Health & Safety Executive, guidelines on safety laws. www.hse.gov.uk.

11. *Latex and You*, a Health & Safety Executive leaflet on latex and its associated health problems. www.hse.gov.uk/pubns/indg320.pdf.

12. Information on health effects of PVC from Ecology Centre website www.ecologycentre.org, the sources for which are: Centers for Disease Control Report, *National Report on Human Exposure to Environmental Chemicals*, 2001; Dadd, Debra, *Home Safe Home*, Penguin Putnam, New York, 1997; Ecology Center Plastic Task Force Report, Berkeley, CA, 1996; Goettlich, Paul, *What are Endocrine Disruptors?*, 2001; National Resources Defense Council website, *Endocrine Disruptors FAQ*, 2001, www.nrdc.org.

13. See 11 above.

14. Greenpeace report on phasing out PVC. www.greenpeace.org.uk/toxics/phasing-out-pvc.

15. Four-year follow-up of effects of toluene diisocyanate exposure on the respiratory system in polyurethane foam manufacturing workers, Journal of Occupational and Environmental Health, April 1992.

16. The Foundation for the Study of Infant Deaths guidelines on second-hand mattresses, based on a study, recommends as follows: "FSID has, as of 6 September 2005, changed its advice on used mattresses as a result of recent research – several studies by Jenkins *et al* show that mattresses can become colonised by bacteria which can stay in them, particularly in the foam interior, a very long time. Also, epidemiological research in Scotland has found that there is a link between using a second-hand mattress and a higher incidence of cot death."

17. See 16 above.

18. See 2 above.

19. Government guidelines on car seat safety for children under 11 years old: www.thinkroadsafety.gov.uk/campaigns/childcarseats/childcarseats.htm.

20. See 19 above.

Chapter 2: Gifts and Toys

1. For more information on these vouchers and on Child Tax Credit see www.childtrustfund.gov.uk, 0845 302 1470, and the Government's website on Child Trust Funds, www.hmrc.gov.uk, 0845 302 1444, for general tax and revenue queries.

2. From article 'Tarnishing The Earth: Gold Mining's Dirty Secret' by Scott Fields in *Environmental Health Perspectives*, www.ehponline.org. *Environmental Health Perspectives* (EHP) is a monthly journal of peer-reviewed research and news on the impact of the environment on human health, published by the National Institute of Environmental Health Sciences.

3. World Health Organisation figures, taken from the above article.

Chapter 3: Nappies

1. EDANA, the international association which represents the nonwovens and hygienic products industry. www.hapco.edana.org.

2. Mother-ease research: "Another chemical that simply should not be in a baby's diaper [nappy] is sodium polyacrylate. This substance is found in the fluff layer of the disposable and turns your baby's urine into gel. Sodium polyacrylate can absorb 100 times its weight in liquid. It makes for a very absorbent diaper, but has been linked to Toxic Shock Syndrome in tampon use. In the past, use of this chemical has been associated with severe diaper rash and bleeding perineal and scrotal tissue, because it pulls fluid so strongly that it excoriates human tissue. No neutral long term study of any kind has been done to assess the affect over time, of contact of this substance with vulnerable genital tissue." See www.mother-ease.com.

3. List of ingredients from Huggies website, www.huggies.com. "The inside absorbent padding on Huggies® nappies is made of wood cellulose fibre, a fluffy paper-like material, and a super-absorbent material called polyacrylate. Other materials used include polypropylene, polyester, and polyethylene. These are all synthetic materials designed to enhance the fit of the nappy and to help stop leaks. The elastic strands in all Huggies nappies are made of synthetic rubber to provide a snug but gentle fit for baby. In addition, Huggies nappies feature an all-over breathable outer cover." Though promoted as positive points, the list of plastic and synthetic ingredients is hardly reassuring if you want to keep your baby away from these products.

4. Source: *Pediatrics,* Vol. 116, September 2005.

5. Source: article by Lauren Alberta, Susan M. Sweeney and Karen Wiss in *Pediatrics,* Vol 116, September 2005, pp. e150-e152.

6. Women's Environmental Network, www.wen.org.uk.

7. Women's Environmental Network, as above.

8. Costs from article by Lucy Jewson, on the website of Cut 4 Cloth, www.cut4cloth.co.uk.

9. Source: Sept/Oct 99 issue of *Archives of Environmental Health.*

10. Archives of Disease in Childhood, researchers from the University of Kiel.

11. Women's Environmental Network, as above.

12. Women's Environmental Network, as above.

13. Women's Environmental Network, as above.

14. See National Association of Nappy Services, www.changeanappy.co.uk/questions.htm. 0121 693 4949.

Chapter 4: The Nursery

1. From Self-Coat Eco Paints, www.self-coat.co.uk. "A VOC is defined as a volatile organic compound. These are emitted as gases from certain solids or liquids. VOCs are widely used as ingredients in household products, such as paints, varnishes, wax and cleaning products. These products release VOCs while you are storing them or using them and remain in the environment for some time afterwards. VOCs can cause eye and respiratory tract inflammation, headaches, dizziness, nausea and vomiting, tiredness, skin irritation and allergic reactions. There is also evidence that cancers and mutations can be caused."

2. 'Multi-system disorder after exposure to paint stripper (Nitromors)', by N. A. Memon and A. R. Davidson, *British Medical Journal,* March 2001.

3. Product Review from *Environmental Building News*, November 1997, on website Building Green: www.buildinggreen.com/auth/article.cfm?fileName=061005c.xml.

4. 'Tree Hugger: Electric Smog; More on Frequency Pollution' which has reference to other articles and websites.
www.treehugger.com/files/2006/03/electrical_smog.php.

Chapter 5: Clothes

1. Fairtrade Foundation. www.fairtrade.org.uk. Tel 020 7405 5942.

Chapter 6: Food

1. Food Standards Agency. www.food.gov.uk.

2. AgriStats, a website containing statistics from DEFRA. www.ukagriculture.com.

3. For articles on the benefits of organic eating, see the Soil Association website: www.soilassociation.org. Tel 0117 314 5000.

4. Comment on Food Commission report by David Thomas, researcher, nutrition and mineral adviser, writing in *Food Magazine*, Jan/Mar2006. www.foodcomm.org.uk/PDF%20files/meat_dairy2.pdf.

5. 'Organic Diets Significantly Lower Children's Dietary Exposure to Organophosphorus Pesticides', by Chensheng Lu *et al.*, Department of Environmental and Occupational Health, Rollins School of Public Health, Emory University, Atlanta, Georgia, USA.

6. Department of Environmental and Occupational Health Sciences, University of Washington, Seattle, Washington, USA. www.depts.washington.edu/envhlth/topic/children.php; National Center for Environmental Health, Centers for Disease Control and Prevention, Atlanta, Georgia, USA. Report published in Environmental Health Perspectives, Volume 114 No. 2, February 2006. (www.ehponline.org/docs/2005/8418/abstract.html).

7. National Childbirth Trust (www.nct.org.uk) Breast-feeding Line 0870 444 8708, Pregnancy and Birth Line 0870 444 8709.

8. British Dietetic Association: Paediatric Group Position Statement on the use of Soya Protein for Infants (www.bda.uk.com/downloads). For more advice and research on infant formulae, see the Department of Health (www.dh.gov.uk/en/Healthcare/Maternity/Maternalandinfantnutrition/DH_4099) and on same site 'Update on infant formula legislation'. Also see the Food Standards Agency website for 'Tighter controls on baby milk' article (www.efsa.europa.eu/EFSA/efsa_locale-1178620753812_1178620767562.htm).

9. Compassion in World Farming, www.ciwf.org.uk, 01483 521950.

10. Consumer Concerns about Hormones in Food, prepared by Renu Gandhi and Suzanne M. Snedeker for Cornell University USA's College of Veterinary Medicine (envirocancer.cornell.edu/Factsheet/Diet/fs37.hormones.cfm).

11. 'Bovine Spongiform Encephalopathy: Current Status and Possible Impact' by J. Eric Hillerton, Institute for Animal Health, Compton, Newbury, Berks, United Kingdom RG20 7NN. Reported in *Journal of Dairy Science*, www.jds.fass.org.

12. 'Persistent Environmental Contaminants in Fish and Wildlife', by C.J. Schmitt and C.M. Bunck, National Biological Service for US Department of the Interior Biological Service, http://biology.usgs.gov/s+t/noframe/u208.htm.

13. Viva Vegetarians International Voice for Animals, www.viva.org.uk, 0117 944 1000.

14. Soil Association GM page, www.soilassociation.org/web/sa/saweb.nsf/GetInvolved/geneng.html.

15. GM: the facts, see Soil Association as above. Also see GM Food News, www.gmfoodnews.com.

16. Food Standards Agency (www.food.gov.uk/gmfoods/gm_labelling).

17. Soil Association, as above.

18. Soil Association, as above. Friends of the Earth has regular updates on the safety of our food. GM Freeze run a campaign to ask the Government to put a freeze on the use and import of GMs: see www.gmfreeze.org.

Chapter 7: Inside the House

1. Grist Environmental News & Commentary: report on toxins in the home, www.grist.org/advice/possessions/2003/03/18/possessions-cleaning/index.html.

2. Greenpeace: 'Poisoning the unborn survey' September2005, www.greenpeace.org.

3. 'How Dangerous Are Household Cleaners? More Than 7 Million Accidental Poisonings Occur Each Year', Kathy Browning, 2006. Associated Content (www.associatedcontent.com). Report in *New Scientist*. www.grist.org/advice/possessions/2003/03/18/possessions-cleaning/index.html).

4. Figures from Ecozone, www.ecozone.co.uk.

5. Each year around 4,400 children under five are hospitalised following accidents with household washing and cleaning products. Figures from Royal Society for the Prevention of Accidents (RoSPA), www.rospa.com.

6. List of harmful chemicals from *Clean House, Clean Planet* by Karen Logan, Pocket Books, 1997.

Chapter 8: Toiletries

1. Greenpeace Report: 'Poisoning the unborn survey', September 2005, www.greenpeace.org.

2. Herlofson, Bente and Pal Barkvoll, 'The effect of two toothpaste detergents on the frequency of recurrent aphthous ulcers', *Acta Odontol Scand* 1996; 54(3): 150-153.

3. Women's Environmental Network, www.wen.org.uk/cosmetics/facts.htm. 'How Absorbing'.

4. Greenpeace, as above.

5. *Consumer Europe 2002/3*, 18th Edition, Euromonitor International PLC, 60/61 Britton Street, London EC1 5UX.

6. The Cosmetic Site, www.thecosmeticsite.com/news/news_industry.html.

7. Women's Environmental Network. In a random check, WEN found preservatives suspected of mimicking the female hormone oestrogen in 57% of products – this is especially worrying for women when lifetime increased exposure to oestrogen is linked to a heightened risk of breast cancer (WEN research). One preservative, propyl paraben, has been shown to adversely affect male reproductive functions: at the 'daily intake level' currently acceptable under EC law, it decreased daily sperm production. (See Oishi, S., 'Effects of propyl paraben on the male reproductive system', *Food and Chemical Toxicology*, 40: 1807-1813 (2002). Also, Health Report website, with article on toiletries, www.health-report.co.uk. See also the campaign for safe cosmetics, www.safecosmetics.org.

8. WEN, as above.

9. Campaign for Safe Cosmetics, www.safecosmetics.org or www.safecosmetics.org.uk.

10. Greenpeace, as above.

11. Phthalates in baby lotion, shampoo and powder: 'Baby Care Products: Possible Sources of Infant Phthalate Exposure', Sheela Sathyanarayana *et al.*, *Pediatrics*, Vol. 121 No. 2, February 2008, pp. e260-e268.

Resources

SUPPLIERS AND ORGANISATIONS

Commonly used terms

1. Carbon footprint
www.direct.gov.uk/actonco2 is the Government's website on reducing CO_2. See also www.carboncalculator.co.uk and www.energysavingtrust.org.uk.

2. Fair-trade
The Fairtrade Foundation is the official body in the UK for the accreditation of products given the 'Fairtrade' label. This is found on food and items that meet their stringent criteria. www.fairtrade.net is the website of the international Fair Trade labelling scheme. www.fairtrade.org.uk, 020 7405 5942.

Chapter 1: Getting Ready for your Eco Baby

1. Organic and fair-trade cotton bibs
Babi Pur, www.babipur.co.uk, 01766 515346.
Ethical Babe, www.ethicalbabe.com, 0870 043 4821.
Fair and Fabulous, www.fairandfabulous.co.uk, 01689 840792.
Green Baby, www.greenbaby.co.uk, 0870 240 6894.
Lejurra, www.lejurra.co.uk, 0114 272 3060.
The Natural Store, www.thenaturalstore.co.uk, 01273 746781.
Spirit of Nature, www.spiritofnature.co.uk, 0870 725 9885.

2. Glass feeding bottles
Born Free, www.babybornfree.co.uk.

3. Ethical cutlery
Green Tulip, www.greentulip.co.uk, 01380 818 515.

4. Organic bedding
A Lot of Organics, www.alotoforganics.co.uk.
Green Baby, www.greenbaby.co.uk.
Spirit of Nature, www.spiritofnature.co.uk.

5. Organic mattresses
Abaca, www.abacaorganic.co.uk, 01269 598491.
By Nature, www.bynature.co.uk, 020 8488 3556.
Graig Farm Organics, www.graigfarm.co.uk, 01597 851655.
The Healthy House, www.healthyhouse.co.uk, 0845 450 5950.
Natural Mat, www.naturalmat.co.uk, 020 7985 0474.

6. Dust mite-proof cotton mattress covers
The Healthy House (as above).

Chapter 2: Gifts and Toys

1. The Woodland Trust campaigns for the protection of ancient woodland and trees across the UK. Ancient woodland (land that has been wooded for at least 400 years) is our richest habitat for wildlife, and is irreplaceable. Ancient trees also have great cultural, historical and ecological significance. You can help preserve these woodlands and contribute to their many successes so far.
www.woodland-trust.org.uk, 0800 0269650.

2. The Family Tree from John Lewis, www.johnlewis.co.uk, 0845 604 9049.

3. Organic search engines
A Lot Of Organics, www.alotoforganics.co.uk.
Ethical Junction, www.ethical-junction.org.

4. The Fairtrade Foundation – see *Commonly used terms* (2), p.125.

5. Child sponsorship
Action Aid UK, www.actionaid.org.uk, 01460 238000.
Oxfam, www.oxfam.org.uk, 0870 333 2700.
Plan International, www.plan-international.org.uk, 020 7482 9777.
Save the Children, www.savethechildren.org.uk, 020 7012 6400.
World Vision UK, www.worldvision.org.uk, 01908 841000.

6. Gift ideas to help communities
Action Aid, www.giftsinaction.org.uk, 01460 238000.
Oxfam, www.oxfam.org.uk, 0870 333 2700.
Practical Presents, www.practicalpresents.org, 01926 634400.
UNICEF, www.unicef.org.uk, 0870 075 4000.

7. London Zoo and Whipsnade Wild Animal Park
www.zsl.org/london-zoo. This website has links to several animal charities including the World Wildlife Fund and the RSPCA.
www.donation4charity.org/wwf-adoptions.php.

8. Key child tax credit providers

The Children's Mutual offers ethical accounts and accounts operated according to Islamic/Sharia law. www.thechildrensmutual.co.uk. 0845 077 1899 (Child Trust Fund) or 0845 608 0045 (Sharia Baby Bond).

The Co-operative Bank, www.co-operativebank.co.uk, 08457 212 212.

Evolve Financial Planning, www.evolvefp.com, 020 7965 4700.

Family Investments (trading name of Family Equity Plan Limited), www.familyinvestments.co.uk, 0800 616695.

Methodist Chapel Aid Ltd offers ethical policies, www.methodistchapel.co.uk, 01904 622150.

9. Other ethical funds

Marks & Spencer has an Ethical Fund that invests in mostly UK-based companies and avoids any connection with companies that are involved in armaments, the fur trade, gambling, tobacco or pornography and those that conduct or commission animal-testing or the use of child labour. www.marksandspencer.co.uk.
Also see the following:

Ecology Building Society, www.ecologybuildingsociety.co.uk.

Triodos, www.triodos.co.uk.

Family Investments (a mutual organisation), www.family.co.uk/investments.asp#ectf.

10. Financial advisers

Barchester Green Investors, www.barchestergreen.co.uk.

Ethical Investors Group, www.ethicalinvestors.co.uk.

The Gaeia Partnership, www.gaeia.co.uk, 01342 826703.

To explore and research the ethics and green credentials of other financial institutions, look at:

Ethical Consumer, www.ethicalconsumer.org.

Ethical Investment Research Services (EIRIS), www.eiris.org.

UK Social Investment Forum, www.uksif.org.

11. Green Karat www.greenkarat.com.

12. British Jewellers' Association www.bja.org.uk, 0121 237 1110.

13. Silver Chilli A fair-trade company, they pay their jewellers, who live and work in small villages in the Mexican countryside, 50% of the price of the items up front, so that the producers are not out of pocket, then the balance on delivery, so they are never kept waiting for their money. Most of the profits are then reinvested back into the community. www.silverchilli.com.

14. More ethical jewellery

Adili, www.adili.com, 01258 837437.

Fairwind Trading, www.fairwindonline.com, 020 8374 6254.

Get Ethical, www.getethical.com.

Natural Collection sells a variety of ethical products and supports a large range of not-for-profit organisations and charities including Greenpeace, Friends of the

Earth, the Vegetarian Society and the World Society for the Protection of Animals. www.naturalcollection.com, 0845 3677 001.

The Natural Store, see Chapter 1 Note 1 (p.125).

Tearcraft jewellery and accessories, www.tearcraft.org, 0870 220 323.

15. Wooden toys

Toys To You has a range of eco-friendly wooden mobiles with themes like the farmyard, Noah's Ark and teddies, made from sustainable woods and painted in bright colours. www.toys-to-you.co.uk, 0870 760 7217.

Spirit of Nature, see Chapter 1 Note 1 (p.125).

16. Organic toys

Green Baby, www.greenbaby.co.uk, 0870 240 6894.

People Tree, www.peopletree.co.uk, 0845 450 4595.

So Organic, www.soorganic.com, 0800 169 2579.

17. Fair-trade toys

Minka is a non-profit organisation working with native communities in the poorest regions of Peru. A major objective is to ensure that a fair price is paid for traditional hand-knitted garments. It then in turn provides support such as training for producer groups in many areas, including hand-knitters around Juliaca, ceramic producers from Puno, and a group which makes musical instruments. Their products are sold on **Traidcraft**, www.traidcraft.co.uk, and **Ethical Superstore**, www.ethicalsuperstore.com.

Kenana Knitters started in 1998 in Njoro, Kenya, the primary object being to help rural women find some much-needed income, utilising their spinning and knitting skills. The group buys homespun wool produced locally, which is dyed with natural plant and vegetable dyes, then knits it into toys, clothing and other accessories. Each toy varies in appearance and is truly unique – what's more, it's signed by the person making it! www.kenanaknitters.com.

Global Exchange has come up with a great idea – for each of their fairly traded toys you buy, they will donate one to a needy child in Zimbabwe. They have an adorable mummy doll complete with baby that sits in a pouch on her back, made by members of a craft project who are mothers of disabled children living in Dzivarasekwa township in Zimbabwe. Also stocked are Eco Toy Animals from Sri Lanka, made from handwoven fabrics and dyed in Sri Lanka using environmentally safe pigments. Choose from a variety of colourful critters: Elephant, Crocodile, Horse, Camel, Giraffe, Rhinoceros, Platypus or Turtle. Global Exchange also offers perhaps the softest teddies in the world, made from alpaca wool. www.globalexchange.org.uk .

Toys to You as above (Note 15).

Natural Collection – see Chapter 2 Note 15.

18. To find more ethically produced toys

Clean Up Fashion, www.cleanupfashion.co.uk.

Ethical Trade, www.ethicaltrade.org.

Ethical Junction, www.ethical-junction.org.
Green Search, www.greensearch.co.uk.
Living Ethically, www.livingethically.co.uk.

19. Wiggly Wigglers www.wigglywigglers.co.uk.

Chapter 3: Nappies

1. The Real Nappy Association www.realnappy.com, 01983 401959.

2. The Nappy Lady www.thenappylady.co.uk, 0845 652 6532.

3. Organic cotton terry squares
Try these suppliers and advisers:
Babeco, www.babeco.co.uk, 0117 941 4839.
Cambridge Baby, www.cambridgebaby.co.uk, 01223 572228.
Makes a Change, www.makesachange.co.uk, 07841 070580.
The Nappy Lady, www.thenappylady.co.uk, 0845 652 6532.
The Washable Nappy Company, www.thewashablenappy.co.uk, 020 8376 0307.

4. Popular washable nappy brands – both one-part and two-part. Some companies offer both. Note that many of these companies stock more than one brand, so check their stocks to see what suits you best.
Babeco, www.babeco.co.uk, 0117 935 1609.
Baby Bee Hinds, www.babybeehinds.co.uk, 01253 701518.
Babykind stocks Bimbles, Bumbles, Bamboozle, Fluffles, Wonderoos. www.babykind.co.uk, 0845 094 2275.
Bambino Mio, www.bambinomio.com, 01604 883777.
Bump to 3 www.bumpto3.com, 0870 60 60 276.
Greener Living stocks Diddy Diapers. www.greenerliving.co.uk, 023 9281 8775.
John Lewis stocks Bambino Mio, Kushies and Tots Bots. www.johnlewis.com.
Kitty Kins stocks Fuzzi Bunz, Ecobotts, Popolino. www.kittykins.co.uk, 01986 784445.
Little Green Earthlets, www.earthlets.co.uk, 0845 072 4462.
Motherbliss stocks Kushies, Magic-alls, Motherease. www.motherbliss.com, 020 8925 6150.
Nature Babies, www.naturebabies.co.uk, 01509 621879.
Onelife, www.onelifeworld.co.uk, 01736 799512.
Sam-I-Am, www.sam-i-am.co.uk, 0845 370 8926.
Snazzy Pants, www.snazzypants.co.uk, 0845 370 8440.
Tots Bots, www.totsbots.com, 0141 778 7486.

5. National Association of Nappy Services www.changeanappy.co.uk, 0121 693 4949.

6. Eco-friendly disposable nappies
Available from health food shops and through green and eco websites:

Tushies are so far the only eco nappy to use absolutely no gel at all, relying on wood pulp and cotton for absorbency. It is latex-, coloured dye- and perfume-free, does not use any genetically modified ingredients and the woodpulp they contain is 100% chlorine-free. They are made in the US, so bear in mind that they have to be shipped over to the UK. However, there are many more outlets stocking them now so they are being brought over in bulk. They may not be as absorbent as nappies that contain gel, but they score really high on green points .

Tender Care are made by the same company as Tushies but have a small amount of gel for extra absorbency.

Kiddicare, see www.kiddicare.co.uk.

Moltex öko contain minimal amounts of gel. No bleaching agents, perfumes or lotions and so recommended for babies with eczema or sensitive skin. Biodegradable packaging, and outer nappy layer. Contain tea tree extracts for skin care and odour control. Can be composted in around eight weeks make your own compost heap, as they will not turn into compost on landfill sites due to anaerobic conditions. From Ethical Superstore – see Chapter 2 Note 17.

Bambo Nature from Denmark use pulp from a renewable forestry resource and are oxygen-bleached, rather than chlorine-bleached. They use no optical brighteners, perfumes, lotions or moisturisers. The core contains very absorbent starch, which is 100% biodegradable index.

Green Baby, So Organic and **Kiddicare** – see above and Chapter 2 Note 16.

7. Plastic fasteners
Nappy Nippas from **Makes A Change** – see Chapter 3 Note 3.
Cuddlebabes, www.cuddlebabes.co.uk, 01430 425257.

8. Degradable nappy liners
Bambino Mio, www.bambinomio.com for local stockists.
Imseyvimse from **Green Baby** – see Chapter 1 Note 1.

9. Degradable nappy bags
Goodness Direct, www.goodnessdirect.co.uk, 0871 871 6611.

10. Washable nappy bags
Bambino Mio, www.bambinomio.co.uk.
The Nappy Lady, www.thenappylady.co.uk, 0845 652 6532.
Snazzy Pants, www.snazzypants.co.uk.

11. Plastic pants and waterproof covers
Green Baby – see Chapter 1 Note 1.

Chapter 4: The Nursery

1. Natural paints suppliers
Aglaia Paints, www.naturalpaintsonline.co.uk, which has several UK stockists.
Earth Born Paints, www.earthbornpaints.co.uk, 01928 734171 .
Ecos Organic Paints, www.ecosorganicpaints.com, 01524 852371.

Eco Paints, www.ecopaints.co.uk, 0845 345 7725.
Ecotopia, www.ecotopia.co.uk, **0845 094 2181.**
IEKO Paints, www.ieko.co.uk, 01342 824466.
Natural Deco, www.naturaldeco.co.uk.
Self-Coat London Eco Paints, www.self-coat.co.uk, 020 8648 8230.

2. *Varnish, wood stain and other finishes*
Earth Born Paints – see above.
Eco Paints – see above.
Ecos – see above.

3. *Natural wallpaper*
Eco Centric, www.ecocentric.co.uk, 020 7739 3888.

4. *Adhesives*
Earth Born Paints, Volvox Adhesive – see above.

5. *Fume absorbers*
The Healthy House – see *Resources,* Chapter 1 Note 5.

6. *Air purifiers*
The Healthy House – Odasorbers – see *Resources,* Chapter 1 Note 5.

7. *Natural paint-strippers*
The Green Building Store's Homestrip, www.greenbuildingstore.co.uk, 01484 461705.
Natural Collection, www.naturalcollection.com, 0845 3677 001.

8. *Environmentally friendly rugs and carpets*
Ganesha, www.ganesha.co.uk.
Garthenor Organic Pure Wool, www.organicpurewool.co.uk.
A Lot Of Organics (search engine), www.alotoforganics.co.uk.

9. *Environmentally friendly carpet & upholstery cleaners*
Ecozone – Eco-zyme Deep Restore or White Wizard Stain Remover. www.ecozone.com.

10. *Organic curtain fabric by the metre*
Babeco, www.babeco.co.uk, 0117 935 1609.
Fabrics Ltd, www.organiccotton.biz.
Harlands Organic Furnishings, www.organicfurnishings.co.uk, 07984 635726.
The Natural Store – see Chapter 1 Note 1.

11. *Ecozone's BioBulb* – see *Resources,* Chapter 4 Note 9.

12. *Ecozone's Moonlight Nightlight* – see *Resources,* Chapter 4 Note 9.

Chapter 5: Clothes

1. Organic cotton baby clothes
Adili – see Chapter 2 Note 14.
Baby Kind – see Chapter 1 Note 1.
Bellanatura, www.bellanatura.co.uk, 01275 848775.
Cut 4 Cloth, www.cut4cloth.co.uk, 01326 340956.
Get Ethical – see Chapter 2 Note 14.
Green Baby – see Chapter 1 Note 1.
Howies, www.howies.co.uk, 01239 61 41 22.
Hug, www.hug.co.uk, 0845 130 1525.
John Lewis, www.johnlewis.com.
Makes a Change – see Chapter 3 Note 3.
Mini Organic, www.mini-organic.co.uk, 01225 767003.
People Tree – see Chapter 2 Note 16.
So Organic – see Chapter 2 Note 16.
Spirit of Nature – see Chapter 1 Note 1.

2. Bamboo fabric clothes
Bamboo Baby, www.bamboobaby.co.uk.
Bamboo Clothing, www.bambooclothing.co.uk, 020 8940 3958.
Kiddymania, www.kiddymania.co.uk, 07904 350086.
Lejurra – see Chapter 1 Note 1.
The Natural Store – see Chapter 1 Note 1.
Organic Bamboo Clothing, www.organicbambooclothing.com.

3. Hemp fabric
Hemp Fabric UK, www.hempfabric.co.uk, 01271 314812.

4. Organic sleeping bags
Grembo, www.grembo.co.uk.
Green Baby – see Chapter 1 Note 1.

5. Organic, fair-trade and ethical baby clothes
All Green Organics, www.allgreenorganics.com.
Belle & Dean, www.belleanddean.com, 0118 986 9552.
By Nature, www.bynature.co.uk, 0845 456 7689.
Cambridge Baby – see Chapter 3 Note 3.
Ecoboo Limited, www.ecoboo.co.uk, 01325 316202.
Ethical Babe, www.ethicalbabe.com, 0870 043 4821.
Global Kids, www.globalkids.co.uk, 020 8133 8533.
Gossypium, www.gossypium.co.uk, 0800 085 6549.
Greenberry's, www.greenberrys.co.uk, 01787 319985.
GreenFibres, www.greenfibres.com, 01803 868001.
Grembo Organics – see above.
Natural Collection – see Chapter 2 Note 14.
Pure Organics, www.pure-organics.co.uk, 01207 284754.

Schmidt Natural Clothing, www.naturalclothing.co.uk, 0845 345 0498.
So Organic – see Chapter 2 Note 16
The Sounder Sleep Company, www.allergybestbuys.com, 08707 455002.

6. Ethical shoes
Freerangers, www.freerangers.co.uk.
My Vegan Shoes, www.myveganshoes.com.
Vegan Shoes, www.vegan-shoes.com.
Vegetarian Shoe Shop, www.vegetarianshoes.co.uk.

Chapter 6: Food

1. For articles on the benefits of organic eating
See the **Soil Association** website, www.soilassociation.org, 0117 314 5000.

2. Sources of organic baby milk and food
Baby Nat: Health and independent stores, online.
Baby Organix: Most major supermarkets and drug stores, Planet Organic, Booths, As Nature Intended, The Kiddies Kitchen. www.organix.com, 0800 393 511.
Bio Bambini: health food and independent stores, also from Ulula, www.ulula.co.uk, 01362 688060.
Cow & Gate: Supermarkets, pharmacies.
Hipp Organic: Health and independent stores, www.hipp.co.uk. 0845 050 1351.
Holle: Health and independent stores, online.
Mums4: www.mums4.com, 01926 771 285.
Truuuly Scrumptious: www.bathorganicbabyfood.co.uk, 01761 239 300.

3. Soya milk
Farley's: Supermarkets, pharmacies, www.farleyscloserbynature.co.uk, 0800 212 991.
Milupa Prejomin: Supermarkets, pharmacies, www.milupa-aptamil.co.uk.
SMA Wysoy: Supermarkets, pharmacies, www.smanutrition.co.uk, 01628 660633.

4. Food Standards Agency www.eatwell.gov.uk.

5. NHS www.nhsdirect.nhs.uk and go to Common Health Questions on the browser bar.

6. Organic fruit and vegetable box schemes
Look at www.vegboxschemes.co.uk (search engine) to find your local suppliers.

7. Organic meat suppliers
Some organic box schemes deliver meat – see www.vegboxschemes.co.uk (search engine). Most supermarkets have some organic options. Alternatively, find a local butcher that stocks organic meat, or ask your butcher to do so.
Online delivery services:
Abel and Cole – bacon, beef, burgers, ham, lamb, mince, pies, poultry, pork,

sausages, veal. www.abelandcole.co.uk, 08452 62 62 62.
Eversfield Organic Meat Produce – beef, burgers, game, lamb, poultry, pork, sausages. www.eversfieldorganic.co.uk, 0845 601 8004.
Graig Farm Organics – beef, chicken, goat, lamb, pork, mutton, cooked meats, venison, wild boar, sausages. www.graigfarm.co.uk, 01597 851655.
Rother Valley Organics – bacon, beef, burgers, chicken, game, lamb, pies, pork, poultry, sausages, venison. www.rothervalleyorganics.com, 01730 821062.
Sheepdrove Organic Farm – bacon, beef, burgers, duck, gammon, haggis, lamb, mutton, pork, poultry, sausages. www.sheepdrove.com.
Well Hung Meat – beef, sausages, lamb, pork, poultry, veal. www.wellhungmeat.com, 0845 230 3131.

8. Organic and free range poultry

See 7 above. Many supermarkets also now offer free-range organic chickens.

9. Ethical fish suppliers

Marks & Spencer guarantees all its fish is sustainably caught.
Graig Farm Organics www.graigfarm.co.uk.
Abel and Cole www.abelandcole.co.uk.
For an article on which fish is the most ethical choice, see www.bbc.co.uk/nature/animals/features/228what.shtml.

10. Fair-trade food

See www.fairtrade.org.uk for details of stockists of fair-trade food, updates and new products (www.fairtrade.org.uk/products.htm).

Chapter 7: Inside the House

1. Green cleaning

Products include Ecover, Clearspring, Bio D, Biocare.
Bio D, www.biodegradable.biz.
Biocare, www.biocaresolutions.co.uk, 020 7448 5211.
Ecotopia – see Chapter 4 Note 1.
Ecover, www.ecover.co.uk.
Ecozone – see Chapter 4 Note 9.
Ethical Superstore – see Chapter 2 Note 17.
Green Baby – see Chapter 1 Note 1.
The Healthy House – see Chapter 1 Note 5.
Natural Collection – see Chapter 2 Note 14.
Nigel's Eco Store, www.nigelsecostore.com.
Simple Green, www.simplegreen.com.

2. Blocked sinks and drains

Ecozone – see Chapter 4 Note 9 for NEZ1 drain cleaner.

3. Room air fresheners

Ecotopia – see Chapter 4 Note 1 for Eco Mist air freshener.
The Natural Collection – see Chapter 2 Note 14 for Uni Fresh air freshener.

4. Limescale

Ecozone – see Chapter 4 Note 9 for Magno balls and Toilet cistern magnets.
Natural Collection – see Chapter 2 Note 14 for magnetic pipe limescale preventer.

5. E-Cloth www.e-cloth.com, 01892 893131.

6. No-soap cleaning for clothes

Eco Balls, www.ecoballsdirect.co.uk.
Ecotopia – see Chapter 4 Note 1.
Goodness Direct – see Chapter 3 Note 9.
Green Baby – see Chapter 1 Note 1.
The Healthy House for Aquaballs – see Chapter 1 Note 5.
Nigel's Eco Store, www.nigelsecostore.com.

7. Dust mites

The Healthy House – see Chapter 1 Note 5 for all dust mite products.

8. Energy- and water-saving, recycling, environmental

Branch Home, www.branchhome.com.
Centre for Alternative Technology, www.cat.org.uk.
CRed, the website for communities wanting lower carbon emissions, www.cred-uk.org.
Dr. Energy, www.doctorenergy.co.uk.
Earth Huggers, www.earth-huggers.com.
EBICo Green Power Company, www.ebico.co.uk.
Ecocentric, www.ecocentric.co.uk.
Eco Kettle, www.ecokettle.com.
Ecotricity Green Power Company, www.ecotricity.co.uk.
Electrisave, www.electrisave.co.uk.
Greens Tire, www.greenstireonline.co.uk.
Interflush water saving device, www.interflush.co.uk.
National Energy Foundation, www.nef.org.uk.
Orgomans, www.orgomans.co.uk.
Simple Human, www.simplehumanstore.co.uk.
Solar Technology, Solartechnology.co.uk.
Toilet Hippo, www.hippo-the-watersaver.co.uk.
Tree Hugger, www.treehugger.com.

9. Recycle Now www.recyclenow.com.

Chapter 8: Toiletries

1. Natural and organic deodorants

Green People – DeoKrystal, Rosemary Deodorant, Organic Base No Scent and Organic Homme deodorant, www.greenpeople.co.uk, 01403 740350.

Jason – Roll-on or stick deodorants in various fragrances including Aloe Vera, Lavender and Tea, www.jason-natural.com, or buy from local health food shops or online at 01403 790 913.

Lavera – Roll-on or spray in various fragrances, see www.soorganic.com, 0800 169 2579.

Weleda – Spray deodorants in various fragrances, www.weleda.co.uk, 0115 944 8222.

For other natural products:

Absolutely Pure, www.absolutelypure.com, 0870 760 6915.

Botanicals, www.botanicals.co.uk, 01664 464005.

Ethical Babe, www.ethicalbabe.com, 0870 043 4821.

Little Green Earthlets, www.earthlets.co.uk, 0845 072 4462.

My Pure, www.mypure.co.uk, 0845 456 0639.

So Organic – see Chapter 2 Note 16.

2. Essentials oils, extracts etc.

Available from pharmacies, health food shops and some cosmetics shops. Also see

Faith, www.faith.uk.com.

Florame, www.florame.co.uk.

Neal's Yard, www.nealsyardremedies.com.

Phytobotanica, www.phytobotanica.com.

Tisserand, www.tisserand.com.

3. Olive Co-operative

For information on the project, see www.olivecoop.com, 0161 273 1970.

or shop online at www.palestineonlinestore.com.

Aseela Women's Co-operative, www.aseela.com.

4. Baby organic and natural toiletries

Angelique, www.angelique.co.uk, 020 8769 1258.

Aroma Kids, www.aromakids.co.uk, 0845 257 2110.

Babaloo, www.babaloo.co.uk, 020 7720 2422.

Badger, www.badgerbalm.com.

Beaming Baby, www.beamingbaby.co.uk, 0800 0345 672.

Burt's Bees, www.myburtsbees.co.uk, 01227 464076.

Earth Friendly Baby, www.earthfriendlybaby.co.uk, 0845 257 2110.

Earthbound Organics, www.earthbound.co.uk, 01597 851157.

Essential Care, www.essentialcare.co.uk, 01638 716593.

Green Baby, www.greenbaby.co.uk, 0870 240 6894.

Green People – see Chapter 8 Note 1.

Lavera, www.lavera.co.uk, 01557 870 203.

Living Nature, www.livingnature.com, 01794 323 222.

Spiezia, www.spieziaorganics.com, 0870 850 8851.
Weleda, www.weleda.co.uk, 0115 944 8222.

PUBLICATIONS / BOOKS

The Nursery

Books on electrical emissions and health problems
Electrical Hypersensitivity: a Modern Illness by A. & J. Philips.
Mobile Phones and Masts: the Health Risks by A. & J. Philips.
EMF and Microwave Protection for you and your family by A. & J. Philips.
The three books above are all published by EM Fields Publications
(www.emfields.org) and available from Powerwatch, www.powerwatch.org.uk.
Cross Currents: Perils of Electropollution, the Promise of Electromedicine by Robert O.
Becker, published by Tarcher, part of the Penguin group. Available from Amazon.

Food

The Good Fish Guide by Bernadette Clark, Marine Conservation Society.

Diet in pregnancy and for babies
Conception, Pregnancy & Birth by Miriam Stoppard, Dorling Kindersley.
The Organic Baby & Toddler Cookbook by Daphne Lambert and Tanyia Maxted-Frost,
Green Books.
The Complete Organic Pregnancy by Deirdre Dolan and Alexandra Zissu,
HarperCollins.
Your Non-Toxic Pregnancy by Susannah Marriott, Carroll & Brown.

Magazines

Mother & Baby, www.emap.com.
Pregnancy & Birth.
Ethical Consumer, www.EthicalConsumer.org.
The Ecologist, www.theecologist.org.
Green Futures (subscription only), www.greenfutures.org.uk.
Greener Magazine, www.greenermagazine.blogspot.com.
The Green Parent, www.thegreenparent.co.uk.
Healthy and Organic Living, www.healthyandorganicliving.com.
Juno magazine, www.junomagazine.com.
New Consumer (fair-trade magazine), www.newconsumer.org.

The Green Providers Directory search engine lists books and magazines:
www.search-for-me.co.uk.

By Nature search engine lists books and magazines: www.bynature.co.uk.

Index

Also available from Green Books

THE USE-IT-ALL COOKBOOK

100 delicious recipes to make the most of your food

Bish Muir

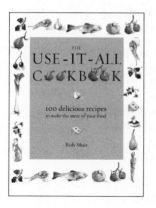

In ***The Use-It-All Cookbook*** Bish Muir offers over 100 recipes and ideas for using your leftovers, and using up that sad-looking carrot or half tub of yoghurt at the back of the fridge. Soups, stews, pies, and risottos sit alongside delicious quick recipes and tasty juices and cakes.

Alongside recipes for turning over 100 left-over ingredients into tasty meals, there is also advice on planning your shopping, storing your food, basic tools for the kitchen, and essential ingredients for the store cupboard; and each recipe features the comparative average cost if bought in the supermarket.

Food waste in landfill sites is a serious contributor to global warming and with changes in people's shopping habits and fears over food hygiene, food wastage is increasing at a rate of 15% every decade. With ***The Use-It-All Cookbook*** you need never throw any food away again, will save money and do your bit for the planet

The Author: Bish Muir has always been interested in food and by helping her mother in the kitchen she learned how to use up all the food in the cupboard, make the most of what ingredients were there and not throw anything away. In ***The Use-It-All Cookbook*** she has called on both her own recipes and those which have been in her family for years. She lives with her husband and children in an increasingly self-sufficient and eco-friendly farmhouse in North Devon.

ISBN 978 1 900322 30 0   £10.95 paperback
To be published in autumn 2008

Also available from Green Books

THE ORGANIC BABY & TODDLER COOKBOOK

From first foods to family meals

Daphne Lambert & Tanyia Maxted-Frost

The Organic Baby & Toddler Cookbook is a comprehensive but easy-to-follow guide to feeding babies from weaning to toddlerhood. It recommends a seasonal, mainly raw or lightly cooked wholefood organic diet, emphasising raw food in spring and summer and lightly cooked foods in autumn and winter.

The cookbook features:

- Why it's so important to feed young children organic food
- An optimum nutrition/organic health guide for breast-feeding mother & child
- Raw juice recipes for weaning baby/breast-feeding mum/toddlers
- The healthy 'grazing smorgasbord' approach to toddler eating
- Daily sprouting and juicing for health
- How to avoid potential health hazards
- Seasonal meal planners & healthy special occasion foods
- Blender salads, toddler savouries, puddings **& much more!**

Daphne Lambert is organic chef/nutritionist of the Penrhos Court organic restaurant and hotel in Herefordshire. She runs regular Pregnancy & Babycare courses at her Penrhos School of Food and Health. **Tanyia Maxted-Frost** has been heavily involved in the organic food scene since 1996. She co-founded the London Organic Food Forum, and created and was Editor of *Organic Food News UK*.

ISBN 978 1 870098 86 1   £6.95 paperback